中英双语

茶知识 100问

100 Questions and Answers about Tea

阮建云　周智修　主编

首批全国优秀出版社　中国农业出版社　农村读物出版社

图书在版编目（CIP）数据

中英双语茶知识100问 / 阮建云，周智修主编. —
北京：中国农业出版社，2022.4
ISBN 978-7-109-27800-4

Ⅰ．①中… Ⅱ．①阮… ②周… Ⅲ．①茶叶－问题解
答－汉、英 Ⅳ．①TS272.5-44

中国版本图书馆CIP数据核字（2021）第019829号

中英双语茶知识100问

ZHONG YING SHUANGYU CHA ZHISHI 100 WEN

中国农业出版社出版
地址：北京市朝阳区麦子店街18号楼
邮编：100125
策划编辑：李　梅
责任编辑：李　梅　　　文字编辑：李　梅　赵世元
版式设计：水长流文化　　责任校对：吴丽婷
印刷：北京中科印刷有限公司
版次：2022年4月第1版
印次：2022年4月北京第1次印刷
发行：新华书店北京发行所
开本：700mm×1000mm　1/16
印张：10
字数：300千字
定价：88.00元

序

　　茶起源于中国，传播于世界。自五世纪开始，中国茶通过陆上和海上丝绸之路传播至世界各地，目前全世界约有30多亿人饮茶，茶丰富了世界各国的物质生活和精神生活。中华茶文化历史悠久、底蕴深厚，是优秀传统文化的重要组成部分，蕴含了"和""敬""清""美""真"的精神内涵。在对外交往中，茶是友谊的桥梁，是和平的使者。

　　中国茶叶学会作为国家级社会团体，多年来积极开展国际交流与合作，是推进中国茶文化国际国内传播的重要社会力量。自2009年起，学会在全国范围内发起"全民饮茶日"活动，已连续成功举办13届。2020年5月21日农业农村部举办首个"国际茶日"活动，学会举办的"美美与共云茶会"作为其分会场之一，邀请到全球18个国家203万茶友在线参与，推动了全民饮茶活动正式走向国际舞台。2020年12月，中国茶叶学会国际茶文化研究与培训中心成立之际，发布了《茶艺职业技能竞赛技术规程》《少儿茶艺等级评价规程》《中国茶艺水平评价规程》《茶叶感官审评水平评价规程》等4项英文版团体标准，这也标志着又一项工作取得了国际化进展。

　　为向全世界传播茶文化和茶科技，亟需创作一批科学、简明的中英双语茶科普读物，因此，中国茶叶学会于2019年启动《中英双语茶知识100问》一书的编译工作。本书内容在历届全民饮茶日编印的《茶知识108问》基础上，精选100个大众热切关注的茶科普问题，组织了17位专家撰写与翻译，开展近

20次的修改和审读，历时2年，用心打磨而成。该书力求科学严谨、考证详实，内容包括茶的起源、茶的制作、茶的饮用、茶的益处、茶的贮藏等5个方面，通过轻快问答的形式、通俗简约的语言来呈现。相信该书必能带领国内外茶叶爱好者走进美妙的茶世界，为弘扬中国茶文化及向全世界传播茶知识尽一份力量！

江用文

Introduction

Tea was grown first in China and later across the world. The Chinese tea, since the 5th century, has been travelling to all parts of the world along the land and maritime Silk Roads. Now the tea-drinking population has reached three billion worldwide, making it an essential vehicle to invite fortunes and happiness. The tea culture in China, which boasts a long and profound history, plays a good part of the country's cultural heritages that include the values of "harmony", "reverence", "integrity", "beauty" and "sincerity". Tea is also considered an envoy of friendship and peace in foreign exchanges.

China Tea Science Society (CTSS) has been for years an active advocate for global exchanges and cooperation to make the Chinese tea culture known to the world community. "National Tea Drinking Day" is a campaign we first launched in 2009, and this year marks the thirteenth event. In celebrating the first "International Tea Day" on May 21, 2020, a theme event of China's Ministry of Agriculture and Rural Affairs, we as a co-sponsor held an on-network Tea Party that had totally 2.03 million tea enthusiasts from 18 countries around the world. The Party therefore marked the premiere of National Tea Drinking activity onto the global stage. The English version of four CTSS standards were announced upon the unveiling of International Tea Culture Research and Training Centre in December, 2020, including *Technical Regulation of Occupational Skills Competition in Tea Ceremony*, *Evaluation Regulation for Level of Teen Tea Ceremony*, *Evaluation Regulation for the Level of Chinese Tea Ceremony* and *Evaluation Regulation for Level of Tea Sensory Evaluation*. This was another effort we made in pursuit of global relations and cooperation.

For worldwide spread of tea science and culture, we feel an urge to provide more bilingual popular readings that should be made clear and comprehensible. The writing of this *100 Questions and Answers about Tea* was triggered therefore in 2019. It basically comes from the previous *108 Questions*, a brochure we prepared for the National Tea Drinking Day. The *100 Questions and Answers*, containing 100 much-concerned tea issues, took two years to complete. It was a project over some 20 revisions that 17 CTSS experts joined to write, translate and proofread. The book itself is rigorous and well-documented, covering the origin, making, drinking, benefits and storage of tea, and is presented in a simple yet flexible language. Dear readers, as we believe, will be exposed to a wonderful picture of teas, and we will, with this book, do our bit for the spread of the Chinese tea culture and the tea science all over the world.

by Jiang Yongwen

前　言

　　为倡导"茶为国饮"，普及茶叶知识，传播茶文化，营造"知茶、爱茶、饮茶"的氛围，弘扬"廉、美、和、敬"的中国茶德思想，自2009年起，中国茶叶学会成功举办了13届"全民饮茶日活动"，在每年4月20日前后开展丰富多彩的全民饮茶科普活动，累计发动600多个城市近千万人参与，已成为参与人数最多、活动范围最广、社会影响最深远的全国性茶事活动。2020年5月21日农业农村部举办首个"国际茶日"，"全民饮茶活动"作为其分会场之一，正式走向了国际舞台。

　　为满足大众对茶科普知识的渴求，中国茶叶学会积极承担传播科学茶知识，弘扬传统茶文化，创作推广优秀茶科普作品的责任。自2015年起，我学会每年组织专家编撰修订茶科普口袋书《茶知识100问》（后更名为《茶知识108问》），至今已编印6册，累计赠阅近60万册，成为目前印量最大的茶知识科普读物，也是历年"全民饮茶活动"的标志性成果，受到了各界的广泛好评，深受全国茶友喜爱。

　　茶作为一个载体，是中华文明的重要名片和独特标识，如今在国际文化交流中扮演着越来越重要的角色。为让全世界认识中国茶，传播科学茶知识，讲好中国茶故事，展现中华传统文化魅力，中国茶叶学会不懈努力，致力于将优秀的茶科普作品推介给全球爱茶的朋友，并为此启动了《中英双语茶知识100问》一书的编译工作，组织国内外17名专家参与撰写、翻译和审读，历时2年，近20次修改和审读，于2021年3月完成了编撰工作。期望通过本双语茶知识读物，向全世界推介健康的中国茶和灿烂的中华茶文化。

本书集科学性、普及性、实用性为一体，分中文、英文两部分，具有四大特色：一是内容系统。本书参考陆羽《茶经》的架构，设计了"茶之源""茶之造""茶之饮""茶之益""茶之藏"等五个部分，精选了100个中外读者最希望了解的茶叶热点高频问题，予以科学严谨的解答。二是编译科学。本书经由多位茶叶专家审阅，确保大量茶学专有名词的正确性，同时又经多位以英文为母语的外籍"茶人"教授翻译和审阅，保证了英文部分的内容适于英语国家读者的阅读习惯。三是形式生动。以"提问—回答"的形式呈现，先启发读者思考，再用简洁生动的语言作答，且中文、英文内容前后排布，保证了阅读的连贯性。四是适于传播。本书言简意赅、论述深入浅出，将专业问题以通俗的语言文字清晰解答，并配以精美图片，轻松易读，美观悦目。

希望《中英双语茶知识100问》的出版能为中华民族灿烂的茶文化走向世界，为普及茶知识、助推茶产业尽微薄之力。

作为学会首本双语茶科普读物，本书编译中难免有疏漏和不足，希望读者批评指正。

• FOREWORD

The thirteen "National Tea Drinking Day" China Tea Science Society (CTSS) have staged since 2009 was a real success to advertize tea as our national beverage, spread tea knowledge and culture, call for a public affection for tea and celebrate the Chinese tea philosophy that emphasizes "integrity, beauty, harmony and reverence". Around the twentieth of April, it was an eventful nationwide celebration that have totally attracted nearly 10 million people from 600-plus cities over the past years, which is considered a national tea-themed campaign that enjoys the largest population, the widest reach over the country and the most far-reaching impact. It was on May 21, 2020 that the "National Tea Drinking Day" premiered on the global stage as a parallel event to the first "International Tea Day" the Ministry of Agriculture and Rural Affairs provided.

CTSS is, to meet public demands, an active advocate for tea knowledge and traditions and an author and publisher of good literary works on tea science. *100 Questions about Tea* (later renamed *108 Questions*), a popular science pocket-sized book CTSS panel of experts wrote in 2015 and annually revised, has totally a circulation of close to 600,000 complimentary copies over six editions to date, which makes it a tea science writing of the largest print-run. The book is also a landmark fruit of the "National Tea Drinking Celebrations" and has established a good reputation among tea enthusiasts nationwide.

Tea, a significant and unique signature of the Chinese civilization, plays an even more striking role in today's global cultural exchanges. CTSS has never ceased efforts to find those tea-themed writings a global access that can help bring the Chinese tea to the world, spread tea science and impress more people with Chinese legends and the culture behind. The preparation and translation of this *100 Questions and Answers about*

Tea (bilingual edition) was therefore launched to this end. It took two years over some 20 revisions before this popular science book was born in March 2021, and 17 CTSS experts from worldwide joined to write, translate and proofread. The book should expectably be an ambassador for the health of the Chinese tea and the splendor of the Chinese tea traditions and culture.

The book tells the popular science and practical use of tea in both Chinese and English. It follows the contents of *The Classic of Tea* (Lu Yu's writing) and gives proven and disciplined answers to a hundred select FAQs we are eager to know in five chapters separately on where tea was from, how to make tea, ways to drink tea, perks of tea and tea collection. The book also provides precise knowledge as it was reviewed by the Chinese experts for correct presentation of tea-related terms, and translated and proofread by several native-speaking tea science professors to adjust the English part to all English-speaking readers. Another feature of the book is the format that places an inspiring question before each answer in short yet animated details. For the integrity of contents, the book features in Chinese for the first half and English for the rest. The book full of gripping descriptions in a clear and popular language, together with select pictures, engages both our mind and eyes.

The unveiling of *100 Questions and Answers about Tea* shall hopefully be a role-player for spreading the Chinese tea culture worldwide, making tea science widely understood and giving a boost for the tea industry.

We understand omissions and errors are inevitable in a book of this type, as the first CTSS bilingual tea science brochure, and any of your comments, suggestions and corrections are greatly appreciated.

目录

茶之源

茶之造

茶之饮

茶之益

茶之藏

100 Questions and Answers about Tea / 75

The Origins of Tea

The Processing of Tea

The Drinking of Tea

The Benefits of Tea

The Storage of Tea

茶之源

茶者，南方之嘉木也。

——唐·陆羽《茶经》

1. 什么是茶树？

　　茶树是中国重要的经济作物，属于常绿木本植物，茶叶边缘有锯齿，叶脉多为7~10对，为网状脉，即叶片主脉明显，侧脉呈≥45度角伸展至叶缘2/3的部位，向上弯曲与上方侧脉相连接，构成网状系统，这是茶树叶片的一个鉴别特征。茶树花一般为白色，种子有硬壳。茶树在植物分类系统中属于被子植物门，双子叶植物纲，山茶目，山茶科，山茶属，茶组。1753年，瑞典植物学家林奈（Carl von Linné）把茶树定名为*Thea sinensis*，意为原产于中国，后又改为*Camellia sinensis*。最后确定的茶树学名为：*Camellia sinensis* (L.) O. Kuntze，沿用至今。茶树的芽叶和嫩梢经加工后就成了茶叶。

茶叶的网状叶脉

2. 茶树有哪些类型？

　　依树型分，茶树有乔木型（主干明显，植株高大）、小乔木型（基部主干明显，植株较高大）和灌木型（无明显主干）三个类型；依叶面积大小分，茶树有特大叶型（叶面积≥

60 cm^2）、大叶型（40 cm^2≤叶面积<60 cm^2）、中叶型（20 cm^2≤叶面积<40 cm^2）和小叶型（叶面积<20 cm^2）四类，叶面积＝叶长×叶宽×0.7。

3. 中国有哪些地方发现了野生大茶树？

中国发现野生大茶树的地方很多。除云南的西双版纳和普洱地区发现有野生大茶树外，云南的其他地区以及贵州、广西、四川、重庆、海南、湖南等地，均发现有高7～26米的野生大茶树。

4. 为什么说中国是茶的故乡？

中国丰富的茶叶史料和现代生物科学技术的鉴定结果，都证明中国是茶的故乡。①中国有最早的关于茶的历史史料记载，如公元前的《诗经》和《尔雅》已有关于茶的记述；汉阳陵出土了全世界最早的茶叶样本，距今已有2100余年；到公元758年左右，唐代陆羽《茶经》更明确记载："茶者，南方之嘉木也，一尺、二尺乃至数十尺，其巴山峡川有两人合抱者。"②中国西南的云贵高原是茶树的起源中心，那里有得天独厚、适于茶树繁衍的自然条件，云贵高原至今尚留存许多古老的野生大茶树。③各种语言中"茶"的读音，都是中国茶字的译音。

另外，茶树原产于中国，传播于世界。当今传布于世界五大洲的茶种、种茶技术、制茶方法、品茶艺术以及茶的文化等，都起源于中国。因此说，中国是茶的故乡。

5. 为什么说茶树的原产地是中国西南地区，有何依据？

　　中国西南的云贵高原，是茶树原产地。其依据是：第一，茶在植物学分类中属于山茶科山茶属，而世界上的山茶科植物主要集中在中国西南地区的云贵高原。山茶科植物共23属380多种，中国西南地区至今已发现15属260多种（近现代在云南、贵州、广西、四川等地考察发现有大量的野生大茶树分布）。第二，目前云贵高原保存有世界上数量最多、树型最大的野生大茶树，这也说明茶树原产于中国西南。第三，根据古地理古气候资料分析，云贵高原部分地区没有受到地壳动态变化的影响，避免了第四纪冰川运动对某些树种的毁灭，被保留下来的古老树种特别多，水杉、银杉、银杏、爪哇紫树、爪哇苦木等被称为"孑遗植物"的第三纪树种。作为热带雨林气候区生长的茶树，亦只有在云贵高原未受到第四纪冰川覆灭的生态环境下，才能生存和繁衍。

云南野生古茶树

6. 中国现代茶区是如何划分的?

中国现代茶区的划分是以自然生态气候条件、产茶历史、茶树类型、品种分布和茶类结构为依据,划分为4大茶区,即华南茶区、西南茶区、江南茶区和江北茶区。

7. 全世界第一部茶学专著是什么?

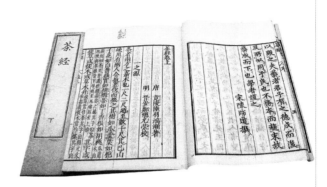

中国第一部也是世界历史上第一部茶学专著是陆羽所著的《茶经》。此书初稿完成于公元8世纪唐代宗永泰元年(765年),经几度修改,定稿于780年。《茶经》全书分三卷十章,共7000余字。其内容为:一之源,二之具,三之造,四之器,五之煮,六之饮,七之事,八之出,九之略,十之图。《茶经》系统地叙述了茶的名称、用字、茶树形态、生长习性、生态环境以及种植要点,介绍了茶叶对人的生理和药理功效,论述了茶叶采摘、制造、烹煮、饮用方法、使用器具、茶叶种类和品质鉴别,搜集了中国古代有关茶事的记载,指出了中唐时期中国茶叶的产地和品质等,是中国历史上第一部茶叶百科全书,也是全世界第一部茶学专著。《茶经》与美国威廉·乌克斯的《茶叶全书》、日本高僧荣西和尚的《吃茶养生记》并称世界三大茶叶经典著作。

8. 中国古代茶圣是谁？

中国古代茶圣是陆羽。陆羽（733—804年）字鸿渐，唐代竟陵（今湖北天门）人。他是个弃婴，由智积禅师抚养，做小和尚，他不愿学佛而喜茶。因安史之乱，陆羽流落湖州，隐居苕溪。数十年中，他深入茶区，考察茶事，躬身实践，总结经验，于唐德宗建中元年（780年）定稿并出版了世界上第一部茶学专著《茶经》。这部专著的问世，有力地促进了茶叶生产和茶文化传播，因而，陆羽被世人称为"茶圣"。

9. 中国茶叶饮用经历了哪几个不同的阶段？

中国饮茶已有几千年的历史。不同的历史时期、不同的茶类有不同的烹饮方法。大体说来，唐代以前，多为粗放煎饮，即将茶叶和姜、盐等混在一起煮，"浑而食之"，叫"茗粥"，也叫"羹饮"。到了唐代，饮用时，先将茶饼烤炙、碾末，然后用水烹煮，称作"煮茶"或"煎茶"。宋代饮用时也是先研茶末，将茶末放在茶盏之中，先加少量水"调膏"，再逐步加沸水并用茶筅击打出泡沫，这种方法叫"点茶"。明代以后主要饮用散茶，饮茶的方法也改为泡饮，即将茶叶置于茶碗或茶壶之中，直接用沸水冲泡。这种方法也叫"撮泡"。泡饮之法，一直沿用到现在。

10. 中国什么时候开始有绿茶？

中国制造绿茶的历史，可以上溯到唐代以前。唐代陆羽《茶经》中所说的饼茶，实际上就是古老的蒸青绿茶。绿茶的加工工艺由晒青到蒸青、炒青、烘青，至创制出片、末、针、眉、螺、珠等形状不同的优质名茶，经历了一个漫长的过程。

11. 中国什么时候开始有红茶？

红茶加工技术源于中国，已有四百多年的历史。据现有文献记载，"红茶"一词最早见于明代刘基的《多能鄙事》一书（15—16世纪）。福建省崇安县（今武夷山市）桐木关首创小种红茶，是历史上最早的一种红茶，因此，崇安县被称为红茶的发源地。1610年，福建崇安产的正山小种红茶首次从海上运往荷兰，然后相继运送至英国、法国和德国等国家。

12. 中国什么时候开始有白茶？

中国古书中就有不少有关白茶的记载，如宋人宋子安的《东溪试茶录》记有："白叶茶……茶叶如纸，民间以为茶瑞，取其第一者为斗茶。"但这只是茶树品种的白茶，而不是加工方法的白茶。后来所谓的白茶是品种与制法相结合的产物。1795年，福建福鼎茶农采摘福鼎大白毫的茶芽，加工成针形茶。1875年，福建发现茶叶茸毛特多的茶树品种，如福鼎大白茶、政和大白茶，1885年起，就用大白茶的嫩芽加工成"白毫银针"。1922年起开始以一芽二叶的嫩梢加工成"白牡丹"。

13. 中国什么时候开始有黑茶？

"黑茶"一词最早出现于明嘉靖三年（1524年）的《明史·食货志》中："……以商茶低劣，悉征黑茶。地产有限，乃

第茶为上中二品，印烙篦上，书商名而考之。每十斤蒸晒一篓，送至茶司，官商对分，官茶易马，商茶给卖。"此时的安化黑茶已经闻名全国，并由"私茶"逐步演变为"官茶"，用以易马。

14. 中国什么时候开始有乌龙茶？

乌龙茶，又称青茶。乌龙茶创制于1725年前后（清雍正年间），福建《安溪县志》记载，安溪人于清雍正三年首先发明乌龙茶做法，以后传入闽北、广东和台湾。另据史料考证，1862年福州即设有经营乌龙茶的茶栈。1866年台湾乌龙茶开始外销。

15. 中国什么时候开始有黄茶？

历史上最早记载的黄茶指的是茶树品种特征，即茶树生长的芽叶自然显露黄色。唐朝享有盛名的安徽寿州黄茶和作为贡茶的四川蒙顶黄芽，都因芽叶自然发黄而得名。

明朝，炒青技术出现后，黄茶的闷黄技术诞生。因为在炒青绿茶的生产过程中，杀青后或揉捻后不及时干燥或干燥程度不足，叶质变黄，但滋味更醇和也更易保存。如黄大茶即创制于明代隆庆年间，距今已有四百多年历史。

16. 中国什么时候开始有花茶？

中国制造花茶已有一千多年的历史。宋时（960年以后）向皇帝进贡的"龙凤饼茶"中虽加入了一种叫作"龙脑"的香料，但这种茶不是严格意义上的花茶。宋代施岳有步月（茉莉）一词，其中写道："玩芳味、春焙旋熏，贮农韵、水沈频爇。"但是否有茶参与焙熏，尚不明了。至元代，文人倪云林有以莲花熏茶的记载。后来，茶中普遍加入"珍茉香草"。明人钱椿年所编的《茶谱》（1539年）一书中所载制茶诸法中，列举有橙茶、莲花茶，并说木樨、茉莉、玫瑰、蔷薇、兰蕙、橘花、栀子、木香、梅花皆可制茶。

17. "白族三道茶"是哪三道？

第一道是苦茶，即雷响茶，把绿茶放在土陶罐里，用文火慢慢地烘烤，并不断地翻抖，待茶叶发出浓香时，即冲入开水，便会发出悦耳的响声。这道茶较苦，饮后可提神醒脑，浑身畅快。

第二道为甜茶，以红糖、乳扇为主料。乳扇是白族的特色食品，是一种乳制品。其做法是：将乳扇烤干捣碎加入红糖，再加核桃仁薄片、芝麻、爆米花等配料，注入茶水冲泡而成。此茶味道甘甜醇香，有滋补的功效。

第三道茶是用生姜、花椒、肉桂粉、松果粉加上蜂蜜，再加入茶水冲泡而成，味麻且辣，口感强烈，令人回味无穷。白族人民用"麻""辣"表示"亲密"，因此"白族三道茶"有着欢迎亲密朋友的意思，是白族同胞接待贵客的礼仪。

18. 什么是酥油茶？

打酥油茶是藏族同胞日常的饮茶方式。酥油茶的原料为茶叶、酥油、盐等。茶叶多为茯茶和砖茶，煮茶的时候将其敲碎。酥油是从牦牛奶、羊奶里提炼出来的，制成块状备用。制作酥油茶时，要先将锅中的水烧开，投入茶块，熬成浓汁，再滤去茶渣，将茶汁倒入专用陶罐内盛放，随时取用。打酥油茶时，先在打茶筒里放好酥油和其他作料，将茶水倒入，盖好茶筒盖，手握一根特制的木杵，上下不停舂打几十下，直到茶汤和酥油充分混合，水乳交融，便是香喷喷的酥油茶了。

19. 谁有"当代茶圣"之誉？

吴觉农有"当代茶圣"之誉。吴觉农先生（1897—1989年）是浙江上虞人，他在青年时代就立志要为振兴祖国农业而奋斗，而他对茶业感情尤深。当他知道中国茶叶历史悠久，曾饮誉世界，之后由于政治腐败，生产落后，茶园荒芜，民不聊生，致使茶业日趋衰退，一蹶不振，世界茶叶市场渐失，于是决心投身祖国茶叶事业。他曾东渡日本，学习现代茶叶科技。留学归国后，即为振兴祖国茶业四处奔波。他与友人合作，拟订了中国茶业复兴计划，先后创办了茶叶出口检验所，建立了茶叶改良场，创建了中国第一个茶叶研究所和第一个培养高级茶叶科技人才的基地——复旦大学茶学系。他还先后到印度、斯里兰卡、印度尼西亚、日本、英国和俄罗斯等地考察访问，以借鉴他国先进经验，探索中国茶业振兴大计。为了茶叶事业，他的足迹遍及全国各地。由于他对中国茶业的贡献巨大，被老一辈无产阶级革命家陆定一誉为"当代茶圣"。

20. 著名茶学家庄晚芳先生倡导的"中国茶德"包含哪些内容？

庄晚芳先生于1989年3月提出和倡导"中国茶德"，其内容为"廉、美、和、敬"。根据庄先生的阐释，"廉"之含义为"清茶一杯，推行清廉，勤俭育德，以茶敬客，以茶代酒"；"美"之含义为"清茶一杯，名品为主，共尝美味，共闻清香，共叙友情，康乐长寿"；"和"之含义为"清茶一杯，德重茶礼，和诚相处，搞好人际关系"；"敬"之含义为"清茶一杯，敬人爱民，助人为乐，器净水甘"。

茶之造

茶之牙者，发于丛薄之上，有三枝、四枝、五枝者，选其中枝颖拔者采焉。

——唐·陆羽《茶经》

21. 现代茶叶如何分类？

茶叶分类的方法有多种，目前较一致的分类法，是将茶分为基本茶类和再加工茶类两个大类。基本茶类按茶叶加工原理和品质特征分为：绿茶、白茶、黄茶、青茶（乌龙茶）和红茶、黑茶，统称六大茶类。再加工茶类主要是指花茶、紧压茶等。

22. 绿茶是如何加工的？

绿茶初加工一般有四道工序，即摊放、杀青、揉捻、干燥。杀青是绿茶加工过程中的关键工艺，其主要目的是利用高温钝化多酚氧化酶的活性，防止茶多酚类物质氧化，从而形成绿茶独特的品质特征。

23. 绿茶如何分类？

绿茶的品质特征是"清汤绿叶"。在中国生产的茶叶中，绿茶是品类最多的一种。以杀青和干燥方法不同，分为炒青、蒸青、炒烘青和晒青四类。目前中国所产绿茶品名极多，仅名优绿茶已达数千种之多。

24. 如何区别炒青与烘青？

两者区别在于绿茶初加工的干燥过程采用方法的不同。所谓炒青，是以炒干的方法干燥，用锅或炒干机作工具，锅下加热，以锅面的接触传热和热的传导蒸发茶叶内的部分水分，达到干燥的目的。所谓烘青，即在干燥过程采用烘干的方法，用烘笼或烘干机作工具，以炭炉或热气发生炉产生热空气蒸发茶叶内的部分水分，达到干燥的目的。炒青绿茶与烘青绿茶品质特征差异为：炒青绿茶条索紧实，色泽绿润，汤色绿明，香气高鲜，有板栗香，滋味浓醇爽口；烘青绿茶与炒青绿茶相比，茶条索松，白毫显露，色泽翠绿，清香带兰花香，滋味鲜醇。

25. 什么季节的绿茶品质较好？

按生产时间，绿茶生产一般分春茶、夏茶和秋茶，一般以春茶质量较好。所以，消费者往往在春季将一年所需的绿茶一次购进，供全年品饮。

绿茶春茶的品质特点是滋味鲜醇。茶树经过一个冬季的营养积累，养分充足，茶叶中有效成分的含量丰富。

随着茶树品种的改进和工艺技术的提升，有些地区夏茶和秋茶的品质也很不错。

26. 西湖龙井茶产于何地？其品质特点如何？

西湖龙井茶产于杭州市西湖区。其品质特点：外形扁平尖削，光滑匀齐，色泽嫩绿匀润；香气鲜嫩清高持久；汤色嫩绿明亮；滋味甘醇鲜爽；叶底嫩匀成朵；有"色绿、香郁、味甘、形美"四绝之美誉。根据国家地理标志证明商标规定的范围，只有西湖产区168平方公里范围内生产的龙井茶，才能称为西湖龙井。

27. 龙井茶有哪几个产区？

根据GB/T 18650—2008《地理标志产品—龙井茶》规定，将龙井茶原产地域划为西湖产区、钱塘产区、越州产区。杭州市西湖区（西湖风景名胜区）现辖行政区域为西湖产区；杭州市萧山、滨江、余杭、富阳、临安、桐庐、建德、淳安等县（市、区）现辖行政区域为钱塘产区；绍兴市绍兴、越城、新昌、嵊州、诸暨等县（市、区）现辖行政区域以及上虞、磐安、东阳、天台等县（市、区）现辖部分乡镇区域为越州产区。

28. 安吉白茶是白茶吗？

安吉白茶的加工工艺流程是绿茶的加工工艺，因此不属于白茶类，是绿茶类。安吉白茶，产于浙江省安吉县，是由叶色白化品种'白叶1号'的鲜叶加工而成，称其为白茶是因为白叶1号为温度敏感型突变体，当春季持续平均气温19～22℃条件下，因叶绿素缺失，茶树萌发的嫩芽为白色。而在高于22℃的条件下，叶色由白逐渐变绿，和一般绿茶茶树一样。安吉白茶是在特定的白化期内采摘、加工和制作的，有叶白脉绿之特点。安吉白茶按加工工艺可分为"龙形"和"凤形"两种，前者用扁形茶加工工艺制作，后者用条形茶加工工艺制作。精品安吉白茶茶条直、显芽，壮实匀整，嫩绿，鲜活泛金边，汤色嫩绿明亮，香气鲜嫩持久，滋味鲜醇甘爽，叶白脉翠，一茶一叶初展，芽长于叶。

29. 缙云黄茶是黄茶吗？

缙云黄茶是用新梢黄化特异茶树品种'中黄2号'，以绿茶加工工艺精制而成，属于绿茶，其氨基酸含量在6.5%以上，茶多酚含量为14.7%～21%，所以口感特别鲜爽，在玻璃杯中冲泡，茶汤清澈，芽叶鲜亮，观赏价值极高。类似的茶还有天台黄茶（由黄化特异品种'中黄1号'新梢加工而成）等。而传统的黄茶是中国六大茶类之一，是通过加工过程中的"闷黄"工序形成，其特点是黄汤黄叶，比如"君山银针"。

30. 红茶是如何加工的？

红茶是以适宜的茶树鲜芽叶为原料，经萎凋、揉捻（切）、发酵、干燥等一系列工艺过程制作而成的茶，其中发酵是红茶加工的关键工艺。红茶发酵的本质是发生了以茶多酚酶促氧化为中心的化学反应，茶多酚（包括EGCG、EGC、ECG、EC等）氧化聚合产生了茶黄素、茶红素等新成分，香气物质比鲜叶明显增加。所以红茶具有红叶红汤和香甜味醇的特征。

31. 红茶的英文名为什么叫"Black tea"？其品质特点如何？

红茶属于全发酵茶，呈现出"红汤红叶"的品质特征，干茶除了芽毫部分显金色，其余以乌黑油亮为好，所以英语中红茶称为"Black tea"。红茶的品质特点是：干茶色泽乌黑油润显金毫，汤色红艳，叶底红亮。中国生产的红茶有工夫红茶、小种红茶和红碎茶三类。这三类红茶各具特色。工夫红茶的特点是外形条索优美，香高味醇；小种红茶的特点是外形肥壮，微带松烟香；红碎茶的特点是叶、碎、片、末分级明显，香味鲜浓。另外，高品质红茶茶汤在冷却后都会有浑浊现象，这种现象称为"冷后浑"，这种浑浊物主要是由咖啡因、茶黄素和茶红素等络合而成。茶汤正常的"冷后浑"现象是红茶品质好的体现。

32. 乌龙茶的工艺特点是什么？

乌龙茶是以具有一定成熟度的茶树鲜叶原料，经过晒青或萎凋、摇青、晾青、杀青、包揉、干燥等工序制出的半发酵茶。摇青和晾青是乌龙茶的特有工艺。

33. 乌龙茶品质特点如何？

乌龙茶主要产于中国福建、广东、台湾三省，按产地不同分为闽北乌龙、闽南乌龙、广东乌龙、台湾乌龙，品质亦有差异。传统乌龙茶品质的共同特点是：外形色泽砂绿、油润；而内质香气高，具有天然的花果香；汤色金黄，滋味浓醇；叶底边缘为红色，中间为绿色，俗称"绿叶红镶边"或称"红边绿玉板"。这个现象的形成是由于乌龙茶制造过程中的做青工序（由摇青和晾青组成），使叶缘碰撞破损红变所致。

34. 武夷岩茶产于何地？

武夷岩茶产于福建省武夷山市所辖行政区范围。山上多岩石，茂密的植被和风化的岩石为茶树提供了丰富的有机质和矿物元素，茶树多生长在山岩之间，故称岩茶。武夷岩茶指在地理标志保护范围内，独特的武夷山自然生态环境下选用适宜的茶树品种进行繁育和栽培，并用独特的传统加工工艺制作而成，具有岩韵（岩骨花香）品质特征的乌龙茶。

岩茶外形条索壮结匀净，色泽砂绿，带蛙皮小白点；内质香气馥郁悠长、具花香，汤色澄黄，滋味醇厚、回味醇爽，独具"岩韵"，叶底肥厚柔软，叶缘朱砂红，叶片中央淡绿泛青，呈绿腹红边。

35. "三坑两涧"是哪里？

武夷山盛产岩茶，且山场众多，每一个山场都有自己独有的小气候。对于常喝岩茶的人来说，"三坑两涧"名气最大。三坑两涧，分别是慧苑坑、牛栏坑、大坑口、流香涧和悟源涧，也是武夷山传统的正岩产区。

36. 大红袍是红茶吗?

大红袍名字中虽有"红"字,却非红茶,而是武夷岩茶中的一种,其加工工艺属于六大茶类中的乌龙茶(青茶)。

大红袍制作仍沿用传统的手工做法,可分为五大工序:萎凋、做青、杀青、揉捻、烘焙。细分为十三道工序,即萎凋—做青(摇青、做手、静置)—炒青—揉捻—复炒—复揉—初焙(走水焙)—扬簸—晾索—拣剔—复焙(足火)—团包—补火。做青工序整个过程中要保持一路香,要达到岩茶传统的三红七绿、绿叶红镶边，偏重偏轻都会影响品质。大红袍条索扭曲、紧结、壮实，色泽青褐油润带宝色，香气馥郁、浓长、幽远之感，滋味浓醇、鲜滑回甘、岩韵明显，杯底余香持久，汤色深橙黄清澈，叶底软亮、匀齐、红边鲜明。

37. 白茶的加工工艺和品质特点是什么？

白茶是中国的六大茶类之一，主产于福建的福鼎、政和、建阳以及松溪等地，属于微发酵茶，按照传统白茶加工工艺（鲜叶－萎凋－干燥）制成。因其制茶原料嫩度和品种不同，白茶可分为白毫银针、白牡丹、贡眉和寿眉等。由于其制作工艺独特，不炒不揉，白毫银针具有外形满身披毫、毫香清鲜、汤色黄绿清澈、滋味鲜醇回甘的品质特点。白茶具有清凉、退热、降火、祛暑的作用和清幽素雅的风格。

38. 黄茶的加工工艺和品质特点是什么？

黄茶是茶叶杀青之后，通过湿热作用堆积闷黄制成的，其品质特点是"黄汤黄叶"，发酵程度也较轻。

39. 黑茶的加工工艺和品质特点是什么？

渥堆是黑茶初制独有的工艺，也是黑茶色、香、味品质形成的关键工艺。黑茶一般原料较粗老，加之制造过程中往往堆积发酵时间较长，因而叶色油黑或黑褐，故称黑茶。品质好的黑茶香气陈香纯，滋味陈醇甘滑，汤色明亮，以橙黄、橙红、红色为佳，叶底黑褐明亮。

40. 什么是普洱茶?

普洱茶是云南的特色茶,是以地理标志保护范围内的云南大叶种晒青绿茶为原料,并在地理标志保护范围内采用特定的加工工艺制成,具有独特品质特征的茶叶。普洱茶按加工工艺及品质特征分为普洱茶生茶和普洱茶熟茶两种类型,按外观形态分为普洱茶(熟茶)散茶、普洱茶(生茶、熟茶)紧压茶。

41. 什么是普洱茶的后发酵?

普洱茶的后发酵,是指云南大叶种晒青毛茶在特定的环境条件下,经微生物、酶、湿热、氧化等综合作用,其内含物质发生一系列转化,而形成普洱茶(熟茶)独有品质特征的过程。

42. 茯茶中的"金花"是什么?

在茯茶加工过程中有一个独特的"发花"工艺,"发花"的实质是以冠突散囊菌为主体的固态发酵过程。人工在茶叶上接种冠突散囊菌,控制温度、湿度等外界条件,以适应它的生长繁殖,该菌形成的黄色闭囊就是"金花"。"发花"工艺对茯茶特有的风味及保健功能的形成具有重要作用。

43. 茶毫多,是好还是不好?

决定茶毫多少的最大因素是品种,有的茶树品种茸毛特别多,有的茶树品种几乎没有茸毛。除了品种外,老嫩也会影响茶毫。一般较嫩的原料,茸毛较多。所以同一品种的茶,一般茸毛多的原料更佳。加工工艺也会影响茶毫，比如龙井的辉锅工艺就是要去掉茶毫，翻炒多的茶毫毛也会掉落得多。

茶毫里含有氨基酸等物质，一定程度上影响着茶叶滋味和营养。茶毫主要是由品种决定的，受原料老嫩和工艺的影响，所以不能根据茶毫多少来判断茶的优劣，有的好茶，原料品种决定没有茶毫。但一般茶毫较多的茶，品质大都不错。

44. 干茶中有微生物吗？

在生活中微生物无处不在，茶叶中当然也有微生物，并且微生物在茶叶生产过程中发挥着重要的作用。例如微生物参与的后发酵是黑茶加工过程中重要的环节。有研究报道，在黑茶中分离鉴定出了多种微生物，包括曲霉、黑霉菌、酵母菌和其他菌类。在这些微生物的作用下，形成了黑茶独特的品质特征。

45. 抹茶是什么茶？

抹茶是采用覆盖栽培的茶树鲜叶，经蒸汽（或热风）杀青后、干燥制成的茶叶为原料，经研磨加工，制成的微粉状茶产品。

46. 抹茶的品质特征如何？

抹茶不仅色翠汤鲜，具有独特的海苔风味，还有氨基酸含量高、叶绿素含量高等特点，且苦涩味低、口感鲜醇，其作为食品添加物在现代食品工业中得到广泛应用。

47. 速溶茶粉是如何加工的?

速溶茶粉是一种能够迅速溶解于水的固体饮料茶,是以成品茶、半成品茶、茶叶副产品或茶鲜叶以及草本植物类、谷物类等为原料,通过提取、过滤、净化、浓缩、干燥等生产工艺,加工制成的一种溶于水而无茶渣的颗粒状、粉末状或小片状的新型饮料。速溶茶粉产品分为速溶纯茶与速溶调配(味)茶两大类,因具有冲饮携带方便、冲水速溶、不留余渣、易于调节浓淡和易于同其他食品调配等特点,越来越广泛地在茶叶市场中推广和应用。

茶之饮

茶有九难：一曰造，二曰别，三曰器，四曰火，五曰水，六曰炙，七曰末，八曰煮，九曰饮。

——唐·陆羽《茶经》

48. 泡好一杯（壶）茶的基本要素是什么？

冲泡一壶好茶，受到各种因素影响，不仅是茶叶的好坏，还受到不同泡茶者的冲泡习惯的影响。有经验的泡茶者了解所泡茶叶的特性，能够根据茶性选择合适的泡法，控制呈味物质浸出速率，突出其品质特点中令人愉悦之处，降低品质缺陷造成的不悦感。泡茶的基本要素可以归纳为以下六个方面：①茶的质量：茶本身质量的优劣是茶汤滋味的先决条件；②冲泡用水：水中含有的矿物元素或其他物质对冲泡结果有影响；③茶器：茶器不同的造型与材质对冲泡结果有影响；④冲泡时间：不同的浸泡时间茶叶中可溶性物质浸出量不同，对茶汤滋味有影响；⑤冲泡水温：茶叶呈味物质和香气成分在不同温度下浸出率和挥发率不同，对冲泡结果有影响；⑥茶水比：不同的茶水比对茶汤浓度有影响。

49. 泡茶时冲泡时间如何把握？

冲泡时间对茶汤滋味有明显的影响，一般泡茶时间从10秒至3分钟都为合理时间范围。在其他因子相同的情况下，冲泡时间越长，茶汤中的水浸出物含量就越高。茶汤中的水浸出物含量和滋味浓度成正相关，因此泡茶时间长，茶汤浓度就高；泡茶时间短，茶汤浓度低。不同人饮茶时喜好的浓度差异较大，有人喜欢喝浓茶，有人喜欢喝淡茶，因此泡茶时间还要根据品饮者喜好而定。

50. 泡茶时水温有什么要求？

一般说来，泡茶水温与茶叶中水浸出物在水中的溶解度呈正相关。水温越高，水浸出物浸出越多，茶汤也越浓。反之，水温越低，水浸出物浸出越少，茶汤也越淡。但不同的茶类对水温的要求是不一样的。冲泡芽叶细嫩的名优绿茶，水温不宜过高，一般以75～85℃为宜，因为名优绿茶的芽叶比较细嫩，用稍低一些的水温冲泡，茶汤嫩绿明亮，滋味也较鲜爽；而在水温较高的情况下，茶叶中的茶多酚类物质容易浸出，影响茶汤滋味，茶叶中所含维生素C也易被破坏。泡茶水温还可根据地域、年龄、性别、习惯等因素进行适当调节。一般来说，冲泡乌龙茶、黑茶、白茶、花茶，则水温要高些，可选用沸水。红茶和黄茶视茶叶嫩度而定，嫩度较高的茶泡茶水温宜低，可选择75～85℃；茶叶嫩度较低则泡茶水温需要较高，可选择85～95℃。

51. 泡茶时茶与水的比例一般是多少？

一般来说，茶多水少则味浓，茶少水多则味淡。如何掌握适度的茶水比例，则要根据茶叶的种类、茶具的大小及饮用者个人品饮习惯来确定。如冲泡名优绿茶、红茶，茶与水的比例大致可以掌握在1:50～1:75，即每杯放3克左右的干茶，加入150～200毫升、75～85℃的水即可。如冲泡普洱茶、乌龙茶，同样的茶杯（壶）和水量，用茶量则应高出一般红、绿茶一倍以上。少数民族嗜好的砖茶，茶汤浓度高，其分解脂肪、帮助消化的功能也强，因此煎煮时，茶和水的比例可以达到1:30～1:40，即50克左右的砖茶，用1500～2000毫升水。

52. 为什么泡茶要茶水分离？

泡茶倡导茶水分离主要有以下几点原因：①茶叶中不同的化学成分浸出规律不同，茶水分离可以通过控制出汤时间使每泡茶品质更均一；②不同的茶类适宜的冲泡次数不一样，茶水分离可以充分体会每泡茶带来的感官享受；③茶水分离会更好地发挥茶的品饮和营养价值，对茶汤色泽、滋味及营养成分的保留更有利。

53. 第一泡茶水是否应该倒掉？

有些人认为第一泡茶不干净，泡茶时总是将第一泡茶水倒掉，认为这样可以洗灰尘、去农残。这种做法是有误的。目前茶叶加工大多实现机械化、连续化、清洁化，这样生产出来的茶叶是很干净卫生的。

市场上销售的茶叶要求必须符合食品安全国家标准，只要消费者购买的茶叶是合格产品，茶叶的卫生指标就值得信赖，可以放心饮用。更为重要的是，茶叶在冲泡第一遍时，大部分氨基酸、咖啡因、维生素C等营养成分已经浸出，将其倒掉就失去了这些营养成分。所以，符合食品安全国家标准的合格茶产品的第一泡不应该倒掉。

54. 不同水质对绿茶茶汤品质有影响吗？

采用不同类型的饮用水冲泡绿茶，会对茶汤品质风格产生较大的影响。日常泡茶用水主要为纯净水（蒸馏水）、天然水（泉水）、天然矿泉水等各类包装饮用水和自来水、水源地水等。通常情况，纯净水（蒸馏水）冲泡的绿茶品质纯正，原汁原味；天然水（泉水）对绿茶香气和滋味品质有一定的改善作用，而天然矿泉水对绿茶品质风格影响较大，多数出现负面影响；自来水因水源地的不同，其影响差异较大，一般大城市的自来水冲泡绿茶品质不佳；碱性水一般不适合冲泡绿茶。一般消费者可选择纯净水（蒸馏水）泡绿茶，而要求较高的消费者可选用低矿化度、低硬度和低碱度的天然水（泉水）。

55. 冲泡绿茶有时会出现白色沉淀是茶叶有问题吗？

在冲泡绿茶时出现白色沉淀主要是水质的问题。茶叶中含一定量的草酸等有机酸，当泡茶用水的硬度较高，即水中含钙和镁离子过多时，就会产生大量的草酸钙、草酸镁等难溶于水的白色沉淀物质，与茶叶的质量关系并不大。

56. 如何使用玻璃杯泡绿茶？

绿茶玻璃杯泡法可分为上投法、中投法、下投法。①上投法是指先在杯中冲入水，然后加入茶叶，此时茶叶在水面缓缓舒展、徐徐下降，这种方法的优势在于不易使茶毫脱落，适合于芽叶细嫩、茸毫含量高易产生毫浑的茶；②中投法是指冲入1/3左右的水，然后加入茶叶，最后注水至七分满，适合于既不过分细嫩、又不十分难以下沉的茶叶；③下投法是指先加入茶叶，然后注水至七分满，适合于不易沉底，芽叶肥壮的茶叶。以龙井茶为例，选用玻璃杯中投法，先在杯中加入100毫升75℃的热水，然后加入3克龙井茶，继续注入150毫升的热水，1分钟后即可。

57. 如何冲泡红茶？

红茶可以选用多种冲泡方法。袋泡红茶和速溶红茶一般采用杯泡法；红碎茶及红茶片、红茶末一般采用壶泡法，便于茶汤与茶渣的分离；工夫红茶和小种红茶可以用盖碗冲泡，也可以采用壶泡法。以工夫红茶为例，选用盖碗冲泡，3克茶用150毫升85℃的水冲泡，第一泡45秒，第二泡20秒，第三泡40秒。外国人饮用红茶，习惯在茶汤中添加牛奶和糖，有的喜欢将茶汁倒入有冰块的容器中，并加入适量蜂蜜和新鲜柠檬，制成清凉的冰红茶。

58. 如何冲泡潮汕工夫茶？

工夫茶流行于福建的闽南地区和广东的潮汕地区，这是一种极为讲究的饮茶方式。喝潮汕工夫茶需用一套古色古香的茶具，人称"烹茶四宝"，一是玉书碨，即一只赭褐色扁形的烧水壶，容量200毫升左右；二是潮汕炉，用以烧开水；三是孟臣罐，一种紫砂茶壶，大小像鹅蛋，容量50多毫升；四是若琛瓯，一种很小的茶杯，只有半个乒乓球大小，仅能容10~20

毫升茶汤。以冲泡凤凰单丛为例，5克茶用100毫升、100℃的水冲泡，第一泡30秒，第二泡20秒，第三泡30秒，此后每泡时间延长10～15秒，一般可以冲泡5～8次。

59. 如何冲泡黄茶？

黄茶一般使用盖碗冲泡，也可选用杯泡或壶泡。黄芽茶、黄小茶等可以参照绿茶冲泡方法和冲泡水温，外形优美细嫩的黄茶可以选用玻璃杯或盖碗冲泡。以莫干黄芽茶为例，3克茶，用150毫升水冲泡，水温80℃，第一泡80秒，第二泡50秒，第三泡60秒。黄大茶需要较高的冲泡水温，一般水温以95℃以上为宜。

60. 如何冲泡白茶?

白茶可选用盖碗下投法冲泡。白茶冲泡水温较高,特别是冲泡白毫银针,需要95℃以上水温,平均冲泡时间较其他茶类稍长;白牡丹、寿眉等干茶叶形较松散,一般选用中至大号盖碗。因白茶汤色颜色较浅,品饮时选用白瓷品茗杯,以便观赏汤色。以白牡丹为例,5克茶,用150毫升水冲泡,选择90℃水温,第一泡60秒,第二泡缩短到30秒,第三泡40秒,第四泡60秒,第五泡80秒。老白茶可以选用大壶煮泡,风味更佳。

61. 如何冲泡黑茶?

冲泡黑茶需要较高温度的水,故常选用大肚紫砂壶冲泡,以保持较高的水温。以普洱茶(紧压茶)为例,5克茶用100毫升水冲泡,水温90℃,第一次冲泡的时间20秒,第二泡缩短到10秒,第三泡延长至15秒,之后每泡延长5～10秒,一般可以冲泡7次以上。

62. 茶叶涩感是怎么形成的?

茶叶中的多酚类物质含有游离羟基,其与口腔黏膜上皮层组织的蛋白质相结合,并凝固成不透水层,这一层薄膜产生一种味感,就是涩味(感)。如果多酚类的羟基很多,形成不透水膜厚,就如同吃了生柿子;如果多酚类的羟基较少,形成的不透水膜薄而不牢固,逐步解离,就形成了先涩后甘的味觉。

63. 为什么有的茶会有青草气？

产生青草气的主要成分是青叶醇，存在于新鲜的茶树叶片中，沸点为156℃。白茶的制作工艺简单，主要是萎凋和烘焙，青叶醇挥发不完全，特别是刚制作出来的当年白茶，常常会带有青草气、青草香。绿茶杀青不足，也会有青草气。

64. 儿童能不能喝茶？

儿童可以喝茶，但是要相对淡些。茶叶里面的咖啡因有提神醒脑效果，若茶水太浓，可能会影响儿童的神经系统，引起过度兴奋，所以建议儿童喝茶要喝淡茶。

65. 隔夜茶能喝吗？

隔夜茶只要未变质，还是可以喝的。但夏季天气炎热，茶汤易变馊，有时上午泡的茶，下午就不能喝了。在人类食物中，有许多含有硝酸盐类的物质，它们在还原酸或在细菌作用下可生成亚硝酸盐，而亚硝酸盐在动物体内的代谢作用下有致癌的风险。由于茶叶中含有一定量的蛋白质，有人便猜测隔夜茶中会有亚硝酸盐的产生，于是便以为喝隔夜茶可能致癌。其实，茶叶中即使产生亚硝酸盐，也是微乎其微的，何况亚硝酸盐本身并不会致癌，它需要一定的条件，即存在二级胺并与之反应生成亚硝胺才有一定毒性。另外，茶叶中含有丰富的茶多酚和维生素C，能抑制亚硝胺的合成。

当然，我们并非提倡人们去饮隔夜茶。任何饮品，大都是以新鲜为好，茶叶也不例外，随泡随饮，不仅香味浓郁，营养物质也更丰富，又可减少杂菌污染。所以，茶最好还是现泡现饮。

66. 可以用茶水服药吗？

因为茶叶中含有咖啡因、可可碱和茶多酚等物质，可能会与某些药物成分发生作用，影响药物疗效，所以服用某些西药时不能饮茶或用茶水送服。例如甲丙氨酯、巴比妥、安定等中枢神经抑制剂就可与茶中咖啡因等兴奋中枢神经因子发生冲突，影响药物的镇静助眠效果；心血管病人或肾炎患者服用潘生丁时若饮茶，茶中咖啡因具有对抗腺苷作用，会减弱潘生丁的药效；贫血病人服用铁剂时若饮茶，茶多酚与铁结合形成沉淀，影响人体对铁剂的吸收。此外，氯丙嗪、氨基比林、阿片全碱、小檗碱、洋地黄、乳酶生、多酶片、胃蛋白酶、硫酸亚铁以及四环素等抗生素药物，都会与茶中茶多酚结合产生不溶性沉淀物，影响药物的吸收。为了充分发挥药物性能，避免出现不良后果，除医生建议需要以茶水送服的药外，服用中西药，最好都不要以茶水送服或吃药后立即饮茶。

67. 西北地区少数民族为什么将茶作为生活必需品？

这与他们的生活习惯及所处地理环境有关。第一，他们日常以乳品、肉食为主要食品，摄入脂肪较多，难以消化，而茶中的内含物有分解脂肪、帮助消化的功能。第二，西北地区海拔高，空气较稀薄，加之气候干燥，人体水分散发较快，需适当的水分补充。饮茶除了能补充水分外，茶叶中的多酚类物质还能刺激唾液分泌，有生津止渴的作用。第三，高原地区，蔬菜水果缺乏，人们易患维生素缺乏症，而茶叶中含有多种维生素，人们可通过饮茶补充维生素。

68. 茶叶入菜的方式有哪些？

茶叶入菜的方式通常有四种，一是将新鲜的茶叶与菜肴一起凉拌、烤制或炒制，是为茶菜；二是在茶汤里加入菜肴一起炖或焖，是为茶汤；三是将茶叶磨成粉撒入菜肴或制作点心，是为茶粉；四是用茶叶的香气熏制食品，是为茶熏。

69. 茶汤到底是酸性还是碱性？

茶汤是酸性的。茶汤的酸碱性取决于茶汤中游离的氢离子和氢氧根离子的相对浓度。茶汤中的酸性物质主要是各种羧酸（柠檬酸、脂肪酸等）、某些氨基酸、维生素C、茶黄素、茶红素等；茶汤中的碱性物质主要是咖啡因和一些香气物质。茶汤的pH与茶的加工、成品茶质量、茶叶的冲泡都有一定关系。通常情况下，所有的茶汤都是酸性的，不存在碱性一说。

茶之益

茶之为用，味至寒，为饮最宜精行俭德之人。若热渴、凝闷、脑疼、目涩、四支烦、百节不舒，聊四五啜，与醍醐、甘露抗衡也。

——唐·陆羽《茶经》

70. 茶叶中的化学成分有哪些?

经过分离鉴定,已知茶叶中的化合物有700多种。茶树鲜叶中,水分占75%~78%,干物质占22%~25%。干物质包括有机物质和无机物质。有机干物质中主要含以下物质:蛋白质20%~30%,糖类20%~25%,茶多酚类10%~25%,脂类8%左右,生物碱3%~5%,游离氨基酸2%~7%,有机酸3%左右,色素1%左右,维生素0.6%~1.0%,芳香物质0.005%~0.03%。

71. 茶叶有哪些功效?

研究表明茶的功效主要有下列23项:①止渴;②缓解疲劳;③舒缓紧张情绪;④强心;⑤利尿;⑥保肝护肝;⑦增强肠道蠕动,缓解便秘,促进排便;⑧消食、解油腻;⑨络合重金属离子(如铅、砷等);⑩抑菌;⑪消炎;⑫抗病毒;⑬抗氧化;⑭抗辐射;⑮预防龋齿;⑯解酒;⑰预防眼科疾病;⑱预防高脂血症,降低血脂;⑲预防糖尿病,降低血糖;⑳预防结石;㉑预防动脉粥样硬化;㉒预防神经退行性病变;㉓预防消化道癌症、乳腺癌、前列腺癌、肺癌等。

72. 茶叶中的茶多酚有哪些功效?

茶多酚主要功效有:①增强毛细血管的作用,增强微血管壁的韧性,效果极为明显;②促进维生素C的吸收,防治维生素缺乏病;③有一定的解毒功效,可将有害金属离子(如六价铬离子)还原成无毒害离子;④抑制动脉粥样硬化,减少高血压和冠心病的发病率；⑤具有显著的体外抗菌杀菌作用；⑥使甲状腺功能亢进恢复正常；⑦抗辐射损伤，提高白细胞数量；⑧抑制突变源引起的突变，抑制癌细胞生成；⑨防止细胞内脂质的过氧化，有抑制自由基生成的作用；⑩有抗凝化瘀作用，降低血脂，防止血栓形成，有减肥功效；⑪预防神经退行性病变，具有预防阿尔兹海默症作用；⑫预防紫外线照射对皮肤的损伤；⑬增强机体免疫力。

73. 茶叶中的生物碱对人体有何作用？

茶叶中的生物碱主要有两种，即咖啡因和可可碱。两种生物碱都属于甲基嘌呤类化合物，是一类重要的生理活性物质，也是茶叶的特征性化学物质之一，均具有显著的兴奋神经中枢、利尿作用。茶叶中的生物碱以咖啡因含量最高，其次为可可碱。

74. 茶叶中的茶氨酸对人体有何功效？

茶氨酸占茶叶中游离氨基酸总量的一半左右，是茶叶中重要的品质成分，尤其与绿茶品质关系密切。茶氨酸具有以下几个方面的功效：①增进记忆力和学习能力；②对帕金森氏症、老年痴呆及传导神经功能紊乱等有预防作用；③保肝护肝作用；④降压安神，改善睡眠；⑤增强人体免疫功能，延缓衰老等。

75. 茶叶中的γ-氨基丁酸对人体有什么功效?

γ-氨基丁酸是一种非蛋白质氨基酸,它广泛地存在于动植物体内。一般情况下茶叶中γ-氨基丁酸含量很低,但通过特殊的加工工艺,其含量能显著地提高。

γ-氨基丁酸具有显著的降血压、改善大脑细胞代谢、增强记忆等效果。还有报道指出,γ-氨基丁酸能改善视觉,降低胆固醇,调节激素分泌,解除氨毒,增进和保护肝脏肝功能等。

76. 饮茶有健齿作用吗?

适量的氟能促使牙齿钙化和牙釉质形成,因此适量饮茶有益于牙齿健康。儿童6岁前是恒牙牙胚形成时期,在这段时期摄取氟元素,对防龋有特殊的作用。儿童一般可以用茶水漱

口的方式补充氟元素。茶叶中多酚类物质的杀菌作用也是健齿作用的主要依据之一。

77. 饮茶具有抗氧化作用吗？

茶叶具有优良的抗氧化活性，这是茶叶具有保健功效的重要基础之一。茶的抗氧化效果与其清除自由基的作用密切相关，茶叶中的多种成分具有清除自由基的功能，其中最主要的成分是茶多酚。茶多酚抗氧化功能的三大机制是：抑制自由基的产生、直接清除自由基以及激活生物体自身的自由基清除体系。因此，饮茶能起到一定抗氧化的作用。

78. 饮茶有防癌的功效吗？

茶叶具有一定的预防癌症的效果。大量动物和人群干预研究都显示，在一定饮茶量的基础上，经常、持续地饮用绿茶可以预防某些癌症，尤其以消化道癌、前列腺癌的预防效应最为显著，肝癌的预防效果也较好。以消化道癌为例，茶叶中的活性成分吸收之后，可以广泛地分布在消化道和代谢器官，从而抑制致癌物质的吸收，并且促进肿瘤细胞的凋亡。茶叶的抗癌作用主要依赖于茶多酚类成分，通过抑制致癌物质（例如亚硝基化合物）、清除自由基、阻断或者抑制肿瘤细胞的生长、促进肿瘤细胞的凋亡从而发挥其防癌的功效。目前，大量的流行病学调查数据表明茶叶在女性群体中的预防癌症作用更为显著，研究发现其原因可能与男性一般都有吸烟等不良生活习惯有关。另外，茶叶中还有丰富的维生素C和维生素E，也具有辅助抗癌功效。

79. 饮茶有助于减肥吗?

人群调查和干预研究发现，茶叶有较好的降脂功效，但是降低体重效果并不突出。经常饮茶可以显著地改善血脂状况，降低血液胆固醇和甘油三酯的水平。中国古代就有关于茶叶减肥功效的记载，如"去腻减肥，轻身换骨""解浓油""久食令人瘦"等。茶叶具有良好的降脂功效是由于它所含的多种有效成分（茶多酚、咖啡因、维生素、氨基酸等）的综合作用。肥胖是因脂肪吸收合成大于分解代谢所引起的。喝茶的过程中，茶多酚可以抑制食物中脂肪的吸收，抑制消化道内酶（糖苷酶、脂肪酶、淀粉酶等）活性，促进肠道蠕动和脂质的排泄，并且增加细胞的线粒体能量消耗，达到减肥降脂的目的。

80. 饮茶有美白作用吗?

茶叶中的茶多酚具有直接阻止紫外线对皮肤损伤的作用,有"紫外线过滤器"之美称。研究表明,茶多酚对紫外线诱导的皮肤损伤有很强的保护作用,抗紫外线的作用强于维生素E。茶多酚还能抑制酪氨酸酶的活性,降低黑色素细胞的代谢强度,减少黑色素的形成,具有皮肤美白的作用。

81. 饮茶可以预防动脉粥样硬化吗?

动脉粥样硬化形成的主要原因是低密度脂蛋白氧化,强烈地抑制巨噬细胞的移动,促使巨噬细胞滞留在动脉壁,导致动脉壁增厚变硬,血管腔狭窄,促进动脉粥样硬化的发生。而茶中的功能成分茶多酚可以调节动脉壁构成细胞的功能,阻碍胆固醇的吸收,抑制低密度脂蛋白氧化,增加高密度脂蛋白的比例。所以喝茶可以预防动脉粥样硬化。

82. 饮茶有防辐射损伤作用吗?

茶叶有防辐射损伤的作用。1962年,苏联学者对小鼠进行体内试验,注射茶叶提取物的小鼠经照射γ射线后,大部分成活,而不注射茶叶提取物的小鼠则大部分死亡,研究还发现,茶叶提取物对造血功能有明显的保护作用。第二次世界大战时,日本广岛原子弹的受害者中,凡长期饮茶的人受辐射损伤的程度较轻,存活率也较高。1973年前后,国内研究证明,用茶叶提取物可防治因辐射损伤而造成的白细胞下降,有利于造血功能的正常化。茶叶抗辐射作用的物质主要有茶多酚、脂多糖、维生素C、维生素E、胱氨酸、半胱氨酸、B族维生素等。尽管茶叶防辐射损伤的机理还有待深入的研究，但很多动物试验和临床试验表明，茶叶防辐射损伤的效果是十分明显的。

83. 饮茶可以预防感冒吗？

饮用白茶可抗炎清火，具有一定预防感冒的功效。①白茶的加工工艺中没有杀青和揉捻，低温长时间萎凋也造成了白茶有别于其他茶类的物质组成，成为其抗炎清火的基础；②白茶中黄酮的含量较高，是天然的抗氧化剂，可起到提高免疫力和保护心血管等作用（一般老白茶的黄酮含量更高）。在感冒初期饮用白茶作用更明显；③白茶功能主要是提高免疫力和预防感冒，但茶终究不是药物的替代品，如炎症和感冒较严重还是建议去医院，毕竟每个人体质以及感冒发生情况是不同的。

84. 饭后用茶水漱口好吗？

饭后，口腔齿隙间常留有各种食物残渣，经口腔内的生物酶、细菌的作用，可能生成蛋白质毒素、亚硝酸盐等致癌物。这些物质可经喝水、进食、咽唾等口腔运动进入消化道，有害健康。饭后用茶水漱口，正好利用茶水中的氟离子和茶多酚抑制齿隙间的细菌生长，而且茶水还有消炎、抑制大肠杆菌、葡萄球菌繁衍的作用。茶水还可将嵌在齿缝中的肉食纤维收缩而离开齿缝。所以饭后用茶水漱口有利健康，尤其是饱食油腻之后。

85. 饮茶有抗衰老的作用吗?

根据衰老自由基学说,老化是自由基产生与清除状态失去平衡的结果,因此减少自由基的生成或对已有自由基进行清除,可有效减慢皮肤的衰老和皱纹的产生。茶叶中的茶多酚是一种抗氧化能力很强的天然抗氧化剂,清除自由基能力超过维生素C和维生素E。所以茶叶能有效预防和减缓皮肤衰老,具有一定美容功效。

86. 如何正确看待茶叶的功效?

虽然茶叶具有很多保健和医用功效,但茶叶毕竟只是一种日常饮用的饮料而不是药物。

对于具有明显器质性损伤,或者经过诊断已经确认某类疾病的人群,如果身体不适,还是应该及时到医院去咨询医生,遵医嘱判断是否可以在治疗或者服药时饮茶。维持身体健康,离不开合理的生活习惯和科学的膳食结构,经常饮茶会有助于健康,降低身体不适发生的概率。

87. 茶汤表面的泡沫是什么?

人们泡茶时经常会看到茶汤表面浮着一层泡沫,产生这种泡沫的物质叫作茶皂素。茶皂素具有很强的水溶性和发泡性,遇水后浸出速率较快,如果再配上热水高冲引起茶叶的翻滚,就会在茶汤表面形成丰富泡沫。茶皂素具有消炎、镇痛的作用,是茶叶重要的功能成分之一。

茶之藏

罗末以合盖贮之，以则置合中，用巨竹刮而屈之，其合以竹节为之，或屈杉以漆之，以纱绢衣之。

——唐·陆羽《茶经》

88. 贮藏中茶叶品质变化主要受哪些因素的影响？

在贮藏过程中，茶叶中的茶多酚、氨基酸、脂类、维生素C、叶绿素等物质极易发生氧化和降解，从而导致茶叶色、香、味等感官品质变化，这种变化主要受茶叶含水量和环境温度、湿度、氧气、光线等因素的影响。茶叶水分含量越高、环境湿度越大、温度越高，茶叶品质改变也越大。在有氧和光线照射下会加速茶叶中多种品质成分的氧化反应，导致茶叶品质变化加快。另外，由于茶叶吸附性较强，极易吸收周围的异味，故贮存茶叶的环境最忌有异味。

89. 家庭如何贮存茶叶？

家庭贮存茶叶主要有三种方法：①容器干燥法。选用体积合适且密封性能好的铁箱、玻璃瓶、陶瓷缸等，底层放入一定量的石灰或硅胶干燥剂，然后将茶叶用牛皮纸包成小包放于干燥剂上，及时更换干燥剂；②小包装密闭干燥法。将茶叶装入密封性能好的塑料复合袋（如铝箔袋）中，加入少量干燥剂并封口装入铁罐或纸罐内；③冰箱冷藏法。利用低温保持茶叶品质稳定的方法既经济又有效。但由于茶叶极易吸附异味和水分，家用冰箱存茶特别要注意包装的阻隔性能，防止茶与其他食物串味。

90. 茶叶受潮后还能饮用吗？

主要根据受潮时间的长短和程度来确定茶叶是否可以饮用。如受潮时间短、影响程度小，茶叶未变质，可立即采取干燥手段（如烘干、炒干等），去除多余水分，茶叶尚能饮用，但感官品质会有影响，如汤色变黄，香气转低。如受潮时间长、影响程度大，茶叶已经变质，甚至霉变，就不能饮用。

91. 不同的茶是否需要不同的存放环境？

不同的茶叶应根据茶叶自身的品质特点选择不同的存放环境。一般不发酵的绿茶和轻发酵的闽南乌龙茶、台湾乌龙茶等以冷藏为佳，品饮前取适量茶叶恢复到室温后即可冲饮；红茶、黄茶、重发酵或焙火的广东乌龙、闽北乌龙等可以采用阴凉、避光、密封、干燥的环境存放，冷藏更佳；白茶、黑茶等需后熟转化的茶叶，可采用棉纸包好后室温避光存放，同时环境湿度不可过高，否则茶叶容易霉变。

92. 茶叶品质易变的原因是什么？

茶叶在贮运和使用过程中，色、香、味品质极易受外界环境条件的影响而发生变化，甚至变质，主要有三大原因：①茶叶结构疏松多孔，具有强吸附性；②茶叶富含的茶多酚类物质极易氧化；③茶叶主要的色、香、味物质容易氧化降解。

93. 名优绿茶应如何贮藏？

名优绿茶贮藏过程中品质极易变化，对贮藏的要求较高。名优绿茶在贮运和使用中主要涉及批量茶贮藏、零售茶叶贮藏和家庭消费贮藏等三个贮藏环节。①批量茶库房贮藏方法：批量茶一般多采用干燥、无异味的库房贮藏和专用冷藏库冷藏。冷藏库的温度一般为2～8℃，相对湿度应小于60%，在库内贮藏8～10个月，茶叶的品质可基本保持不变；②零售茶叶贮藏方法：零售小包装茶应采用阻隔性能好的铝箔复合材料等包装袋，以干燥剂去湿或除氧、抽气充氮等气调技术进行保鲜包装。有条件的也可采用小型冷藏柜冷藏零售茶叶，效果较好；③名优茶家庭贮藏：因数量少，可将茶叶密封后用冰箱或冷柜贮藏，尽量避免和其他带有气味的食品同储。

94. 如何判断茶叶是否劣变？

茶叶品质劣变一般是指茶叶品质发生了较大的变化，并引起了明显的不正常品质问题。一般通过看颜色、闻香气和尝滋味来判断茶叶是否发生了明显的劣变。①看颜色：茶叶外观颜色发生较大的变化，如劣变的绿茶会由翠绿鲜润变黄变暗，红茶外观发暗，好像蒙了一层灰；②闻香气：劣变的茶叶香气浓郁度和纯正度均发生较大的变化，香气低淡，难以持久，甚至出现陈味、油耗味等不令人愉悦的气味；③尝滋味：劣变的茶叶滋味鲜爽度明显下降，滋味品质欠醇正。如劣变的绿茶会失去原有的清鲜，红茶可能会发酸等。

95. 如何识别陈茶与新茶？

陈茶一般指往年生产的茶叶。因贮放时间长，茶叶中的内含物经过长时间的氧化，色香味形等品质均出现较大变化，与当年的新茶存在明显差异。绿茶新茶色泽绿润，有光泽，香高味醇，汤色清明；而陈茶的色泽泛黄，失去新茶固有的新鲜感，香气低，甚至出现陈气，滋味淡、汤色泛黄。红茶新茶色泽乌润，香高味醇，汤色红艳明亮，叶底红亮；陈茶色泽枯暗不润，香气低或有陈气，滋味淡、汤色暗浊，叶底红暗不开展等。

96. 不同茶类的最佳饮用期是多久？

不同的茶叶有不同的最佳饮用时间。不发酵的绿茶和轻发酵的闽南乌龙茶、台湾乌龙茶等一般建议当年饮用；红茶、黄茶、重发酵或焙火的广东乌龙、闽北乌龙等建议在2～3年内饮用完毕；白茶、黑茶等一般需后熟转化的茶叶经过存放能明显提升和改善茶叶品质，建议存放后再行品饮。以普洱茶为例，在适宜的贮藏条件下，一般普洱茶生茶储存15～20年，普洱茶熟茶储存5～8年，品饮口感较好。存茶时间也非越长越好，过长时间的存放，茶叶中的风味物质散失，反而显得平淡，如故宫中的清代贡茶，经过近两百年的存放，香、味俱淡。

97. 普洱茶如何储藏？

普洱茶在存放中应注意含氧量、温度、光线、湿度等因素对茶的影响。①空气：普洱茶仓储要空气流通，但不能放于风口，另外，要注意周围环境有无异味，不可将茶摆放于厨房或其他有生活、工业异味的环境里；②温度：存放普洱茶的温度不可太高或太低，不用人为改变温度，正常的室温就好了，最好常年保持20～30℃，太高的温度会使茶叶加速发酵变酸；③光线，光是一种能量，它能使茶叶内部的化学成分发生变化。如果把茶叶放在日光下晒一天，则茶叶的色泽、滋味都会发生比较显著的变化，从而失去其原有风味，因此，茶叶一定要避光贮藏；④湿度：太干燥的环境会令普洱茶的陈化变得缓慢，如环境较干燥，可以在茶叶旁边摆放一小杯水，令空气中湿度稍微增大。但是太过潮湿的环境会导致普洱茶快速变化，这种变化往往是"霉变"，令普洱茶不可饮用。湿度建议控制在75%左右。沿海一带为海洋性气候，梅雨季湿度会高于75%，更应注意及时开窗通风，降低湿度。

98. 白茶如何存放？

白茶经过存放，茶的颜色变深，茶汤的滋味变醇和，受到很多消费者喜爱。白茶存放有以下要求：①长期储存时，茶的含水率应控制在5%以下；②储存环境应干燥、无异味，湿度控制在60%以下，如库房为水泥地面，底板要用木架支撑；③白茶要低温、避光贮藏，因为在高温、光照条件下，茶叶内含成分的化学变化加快，陈化加速，从而使茶失去原有风味；④储存过程中，没有特殊情况不要打开包装，否则会导致大量空气进入，加速茶的氧化。

99. 普洱茶（生茶）能存放成普洱茶（熟茶）吗？

自然存放条件下，普洱茶（生茶）变不成普洱茶（熟茶）。普洱茶（生茶）是指以云南大叶种晒青毛茶为原料加工制作的紧压茶，干茶色泽墨绿；普洱茶（熟茶）是指以云南大叶种晒青毛茶为原料，经过人工渥堆发酵等工艺加工而成的茶叶，干茶色泽棕褐。所以，普洱茶（熟茶）是经过人工渥堆发酵制成，而普洱茶（生茶）未经过人工渥堆发酵，普洱茶（生茶）在后期贮放过程中会发生品质变化，属于自然陈化，但不会变成普洱茶（熟茶），两者间的风味存在较大的差别。

100. 用于包装茶叶的材料有哪些？

茶叶包装是指从茶叶加工到消费过程中，为确保茶叶的品质和卫生安全，延长茶叶保存期限，便于茶叶的运输、计量、保管、陈列销售及携带所采用的容器或材料。茶叶包装及材料丰富多样，包装用的材料主要有纸板、牛皮纸、白板纸、塑料薄膜、铝箔复合薄膜、玻璃制品等，主要容器包括铁罐、纸罐、铝罐、竹木容器、瓷罐、锡罐等。茶叶包装为保证茶叶在加工、销售、储藏和流通领域的品质，防止和减少茶叶色、香、味和营养成分的变化发挥着重要作用。

100 Questions and
Answers about Tea

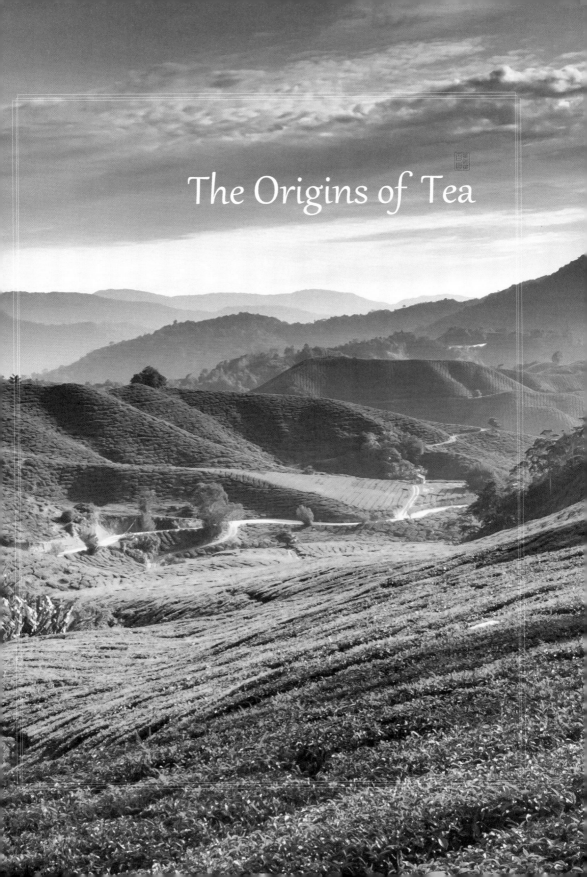

The Origins of Tea

Chapter Information Board

1. What is tea plant?

The tea plant, an important cash crop in China, is an evergreen woody plant. Tea leaves are serrated and each consists of 7-10 pairs of main veins. The lateral veins stretch to two-thirds toward the leaf margin at angles of $\geqslant 45°$, then curve up connecting the upper lateral veins, forming a net. This is a distinguishable characteristic of tea leaves. The color of the flowers is generally white and the seeds have hard shells. In plant taxonomy, tea plant belongs to section *Thea*, genus *Camellia*, family Theaceae, order Theales, class Magnoliopsida, division Magnoliophyta. In 1753, a Swedish botanist Carl von Linné nominated tea as *Thea sinensis*, meaning "tea plant originated from China". It was later renamed as *Camellia sinensis* (L.) O. Kuntze. The buds, leaves and young shoots are harvested (plucked) and processed into made teas.

Tea leaf

2. What are the categories of tea plants?

According to the plant type, tea plants can be divided into three categories: arbor (obvious trunk; large and tall plant), semi-arbor (obvious base trunk; relatively large and tall plant), and shrub (without trunk; small plant). Based on leaf size, tea plants can be divided into 4 types, including extremely large leaf (leaf area $\geqslant$ 60 cm^2), large leaf (40 $cm^2 \leqslant$ leaf area$<$60 cm^2), medium leaf (20 $cm^2 \leqslant$ leaf area$<$40 cm^2), and small leaf (leaf area$<$20 cm^2), leaf area $=$ Length $\times$ Width $\times$ 0.7.

3. Where can the wild ancient tea plants be found in China?

Wild ancient tea plants can be found in many places in China. Some wild tea plants are found in Xishuang Banna and Pu'er in Yunnan Province. Furthermore, some wild tea plants with a height of 7-26 meters have been found in many places such as other places in Yunnan Province, Guizhou Province, Guangxi Zhuang Autonomous Region, Sichuan Province, Chongqing Municipality, Hainan Province, and Hunan Province, etc.

4. Why is China the hometown of tea?

Abundant historical evidence and modern biological science have proved that China is the hometown of tea.

① The earliest historical records of tea are found in China.

For example, *the Odes* and *Erhya*, two ancient (600 B.C.) Chinese books have records of tea. Relic of tea unearthed in Hanyang Mausoleum can be traced back to 2,100 years ago, which might be the earliest sample of made tea in the world. In 758 A.D., Lu Yu of the Tang Dynasty stated clearly in *The Classic of Tea* : "Tea, a brilliant plant in the south, is as tall as one *chi*, two *chi* or even dozens of *chi*; in the gorges and mountain areas of Bashan Mountains, there are tea plants with thick trunks which can only be encircled by two men with their arms joined together."

② There still exists many ancient wild tea plants in the Yunnan-Guizhou Plateau in southwest China.

③ The word for *tea* in different languages around the world inherited the pronunciation from the Chinese word "cha".

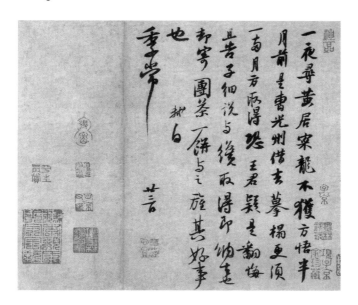

Tea plants are originated from China and spread to the whole world. The Yunnan-Guizhou Plateau in Southwest China is believed to be the center of origin of tea plants. Because of the area's particularly favorable natural conditions suitable for the reproduction of tea plants, there still remains a large number of ancient wild large tea plants. The tea cultivars, tea planting techniques, tea processing, tea tasting art, and tea culture which are spread all over the world have their original marks from China. That's why we say China is the hometown of tea.

5. Why Southwest China is the original center of tea plants? Is there any evidence?

The Yunnan-Guizhou Plateau in China is the original center of the tea plant. The pieces of evidence are as follows. First, tea botanically belongs to the genus *Camellia*, family Theaceae. The Theaceae plants in the world are mostly concentrated in the Yunnan-Guizhou Plateau in Southwest China. The Theaceae family has 23 genus and 380 species, of which 15 genus and 260 species have so far been found in Southwest China. Second, currently Yunnan-Guizhou Plateau boasts the largest number of ancient tea plants in the world, from which we can tell that tea plants originated in Southwest China. Third, paleogeography and paleoclimate show that as the result of dynamic changes in the earth's crust, Yunnan-Guizhou Plateau has once been exempted from the destruction of some types of plants because of the Quaternary Glaciation.

There remain many types of ancient plants, including those "relict plants" in the Tertiary Period, such as metasequoia, cathaya argryophylla, gingko, Java cotton-gum and Java quassia. Originally as a tropical rainforest plant, the tea plant might have survived and flourished only in the ecological environment of Yunnan-Guizhou Plateau against the severe condition of the Quaternary Glaciation.

6. How do we divide the modern tea areas in China?

The division of modern tea areas in China is based on the following factors: ecological and climatic conditions, history of production, types of tea plants, distributions of tea cultivars, and structure of made tea types. Thus, tea production in China is located in four major areas: South China Tea Area, Southwest China Tea Area, South Yangtze River Tea Area, and North Yangtze River Tea Area.

7. Which is the world's first monograph on Tea Science?

The first monograph on tea science in the history of China is *The Classic of Tea* written by Lu Yu. The first draft of this book was finished in the year 765 A.D., the First Year of Yongtai, during the reign of Daizong in the Tang Dynasty. *The Classic of Tea* is divided into three volumes and ten chapters with totally more than 7,000 words. Its contents include: *Chapter 1 The Beginnings of Tea; Chapter 2 The Tools of Tea; Chapter 3 The Manufacture of Tea; Chapter 4 The Équipage; Chapter 5 The Brewing of Tea;*

Chapter 6 Drinking the Tea; Chapter 7 Tea Matters; Chapter 8 The Production of Tea; Chapter 9 Generalities; Chapter 10 Illustrations. It has systematically explained the name of tea, the Chinese character "tea", forms of tea plants, growing habits, growing conditions and key planting knowledge. The book has also introduced the physiological and pharmacological effects of tea, discussed tea picking, processing, brewing, drinking tea, tea wares, identified tea types and quality, collected the records of tea in ancient China, and the production areas and quality of tea in the Middle of the Tang Dynasty in China. It is the first encyclopedia on tea in Chinese history and is also the world's first monograph on tea science. *The Classic of Tea together with All about Tea* written by American author William H. Ukers and *Records of Health-Preserving through Drinking Tea* written by Japanese religious master Yosai are titled the world's three great classic works on tea.

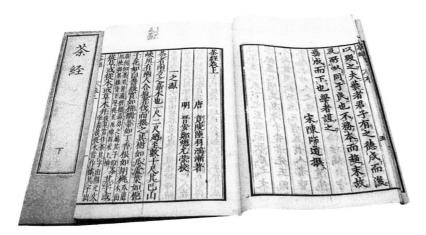

8. Who is the "Sage of Tea" in ancient times?

Lu Yu (733–804 A.D.), whose courtesy name was Hongjian, was a native resident of Jingling in the Tang Dynasty (present-day Tianmen, Hubei Province). He was an abandoned baby and was later fostered by Zen Master Zhiji. He became a little monk but was reluctant to learn Buddhism; instead, he loved tea. During the time of An Lushan Rebellion, Lu Yu was exiled to Huzhou and lived in seclusion in Tiaoxi. In the period of several decades, he visited tea areas to study tea, during which he personally practiced everything related to tea and gained experiences from his practices, and finally he finished and published the world's first monograph on tea, *The Classic of Tea,* in 780 A.D., the first year of Jianzhong, during the reign of Dezong in the Tang Dynasty. The publishing of this monograph has vigorously promoted the production of tea and the spread of tea culture, thus Lu Yu is titled as the "Sage of Tea".

9. What different stages have tea-drinking in China gone through?

Tea-drinking has experienced thousands of years' history in China. Different ways of brewing and drinking were adopted for different types of tea in different historical periods. Generally speaking, by the Tang Dynasty tea was brewed in rough ways which brew tea together with ginger, salt and others, and "take

them all", which is known as "bitter congee" or "soup-like drinks". In the Tang Dynasty, people roasted and ground the tea cake and then boiled it, the process of which was called "boiling the tea" (zhucha) or "decocting the tea" (jiancha). Cake-shaped tea was remained in the Song Dynasty. Just as in the Tang Dynasty, people ground the tea cake into a paste and added them into the tea ware to "prepare the paste", and then added boiling water in a progressive manner and whipped and foamed the tea with a whisk. This is known as "preparing the tea" (diancha). After the Ming Dynasty, people drink mostly loose tea and make tea by brewing instead, which is to place tea in a bowl or teapot and brew with boiling water. This is known as "gather and brew" (cuopao) and has been kept until today.

10. When did the green tea first appear in China?

The history of making green tea in China can be traced back to the time prior to the Tang Dynasty. The cake tea mentioned in *The Classic of Tea* written by Lu Yu is actually the ancient steamed green tea. It took a long time for the processing techniques of the green tea to develop from sun-drying to boiling, frying and baking, and then producing premium famous teas in different forms including blade, powder needle, eyebrow, screw, and bead.

11. When did the black tea first appear in China?

The processing of black tea originated in China and has a history of more than 400 years. According to literature, the word "black tea" first appeared in the book *Capable of Doing All Sorts of Vulgar Things* (15th-16th Century) written by Liu Ji in the Ming Dynasty. The earliest black tea, Souchong black tea, was invented in Tongmuguan in Chong'an County (present-day Wuyishan City), Fujian Province. Thus, Chong'an County is titled the origin of the black tea. In 1610, the Lapsang Souchong black tea produced in Wuyi Mountains was first shipped to the Netherlands by sea, and then later to the United Kingdom, France and Germany successively.

12. When did the white tea first appear in China?

In ancient literature works of China, there are many records about white tea. For example, in the Song Dynasty, Song Zi'an, recorded in *Records of Trying Tea in Dongxi*: "The tea leaf of white tea is like a paper. The folks regarded the white tea as the best amongst all teas, and it was the Number One in the Tea Competition." However, the so-called white tea is only a special variety of tea tree, rather than a different type of made tea with different processing methods. The "white tea" later was the result of both a different variety and different processing methods. In 1795, the tea growers in Fuding City, Fujian Province picked the tea buds of Fuding Dabaicha, and processed the tea buds into "Silver needles"(tea name). In 1875, several varieties of tea plants, for example, Fuding white tea and Zhenghe white tea, with plenty of white hair were found in Fujian Province. Since 1885, the tender buds of these varieties were processed into "Pekoe silver needles". In 1922, the tender tips with one bud and two leaves were processed into "White peony"(tea name).

13. When did the dark tea first appear in China?

The phrase "Dark tea" first appeared in *The History of Ming Dynasty·Records of Food and Goods* written in 1524, the Third Year of Jiajing in the Ming Dynasty: "Due to the poor quality of the commercial tea, the government collects the dark tea. The production of the land is limited, therefore the tea produced is labeled the second-class of medium-high quality, with a thin bamboo strip ironed with a brand, and then is titled "commercial tea". Each ten *jin* (1/2 kilograms) of dark tea is steamed and sun-dried, and then labeled with a thin bamboo strip. The dark tea is then sent to the Tea Bureau for the governmental officials and merchants; the tea given to officials are used to trade for horses, and the other portions to merchants are commercial teas used for sales." At that time, Anhua dark tea has been famous all over the country, and thus has gradually evolved from "privately-owned tea" into "official tea" in order to trade for horses.

14. When did the Oolong tea first appear in China?

The Oolong Tea is also called cyan tea. The Oolong tea was created around 1725 (during the reign of Emperor Yongzheng in the Qing Dynasty). *Records of Anxi County* has recorded that: "Anxi people have first created the processing method of the Oolong tea in the Third Year of Yongzheng. And the Oolong tea was later introduced to the northern Fujian Province, Guangdong Province and Taiwan Province." According to other historical data, in 1862, Fuzhou established teahouses that sold Oolong tea. In 1866, Taiwan Oolong tea started to be sold overseas."

15. When did the yellow tea first appear in China?

Record of the yellow tea in history is taken as the characteristics of tea plants. The color of the growing bud and leaves looks yellow when the tea plants grow up. The most famous yellow teas in the Tang Dynasty were Shouzhou yellow tea from Anhui Province and Mengding yellow tea from Sichuan Province. These teas got such names because of their natural yellow color of the bud. In the Ming Dynasty, with the introduction of the technology of frying green tea in the Ming Dynasty, the heaping and smothering technology came into being. In the process of frying green tea after pan-firing and rolling, when the drying of tea is too late or left unfinished, the color of the tea turns yellow, while the taste becomes more mellow and the tea is conducive to storage. For example, Huangdacha was created during the reign of Rongqing in the Ming Dynasty and has a history of over 400 years.

16. When did the scented tea first appear in China?

China has a history of over 1,000 years in making the scented tea. In the Song Dynasty (after 960 A.D.), the "Longfeng Tea Cake" was added with a spice of "borneol" cmol used to be paid as tribute to the emperors. But this was no true scented tea. Shi Yue, a poet in the Song Dynasty, wrote in his poem named *Roam in the Moonlight (Jasmine)*: "The sweet taste is produced by quick baking

in spring; it embodies the charm of nature as the aroma of sandalwood incense lingers." However, it is still unknown whether the tea was involved in baking. By the Yuan Dynasty, a scholar called Ni Yunlin left a record of lotus-scented tea. Later, it was universal to add "treasured jasmine and other aromatic herbs" into tea. The book *Recipe of Tea* (1539) compiled by Qian Chunnian in the Ming Dynasty has recorded the making methods of multiple teas, enumerating orange tea and lotus tea, and also remarked that osmanthus, jasmine, rose, orchid, orange blossom, cape jasmine, costustoot, plum, etc. can all be made into tea.

17. What are the "Three Teas of the Bai People"?

The first is a bitter tea, i.e., "Thunderous tea". Put the green tea into the earthware jar and bake with a gentle fire, and churn and shake the tea; when the tea give out a special aroma, brew it with boiled water; it will give a melodious sound. This tea is slightly bitter and gives a refreshing feeling after drinking.

The second tea is a sweet tea, featuring brown sugar and *rushan* as the main ingredients. *Rushan*, a specialty food of Bai people, is a dairy product. The making method is: roast the *rushan* till it is dry, mash it, and add brown sugar into it; add ingredients including sliced walnuts, sesame and popcorn; pour tea into the ingredients, thus the tea is made. This tea tastes sweet and full-bodied, and has an efficacy of nourishing our body.

The third tea is made from ginger, pepper, ground cinnamon and ground pinecone, and added with honey and tea. It tastes pungent and spicy with an intense taste, leaving tea drinkers with a long aftertaste. "Pungent and spicy tastes" are used by Bai people to express the meaning of "affinity", thus the "Three Teas of the Bai People" carries the meaning of "intimate friends", which is a ritual for Bai people to receive distinguished guests.

18. What is the buttered tea?

"Pounding" buttered tea is a way of drinking tea daily for Tibetan people. The ingredients of buttered tea include tea, butter and salt, etc. The tea are mainly Fu tea and brick tea. Break the tea bricks into crushes before brewing, and extract butter from Yak milk or goat milk and make butter into blocks. When making buttered tea, first boil the water in the pot, then add the brick tea; stew the tea into a thick soup, and then filter the tea residues. Pour the tea soup into a special jar for later use anytime. When "pounding" buttered tea, first put the butter and other ingredients into a tube, then pour the tea soup into it, and put a cover onto the tube. Hold a wood pestle that sticks out of the tube cover, pound up and down for dozens of times until the tea soup blends with the

butter adequately, i.e., the tea is completely mixed with milk, then we have the aromatic buttered tea.

19. Who deserves the title of "Sage of Tea in Contemporary China"?

Professor Wu Juenong (1897-1989), a native of Shangyu, Zhejiang Province, had been determined to fight for the rejuvenation of China's agriculture in his youth and demonstrated deep emotions for the tea industry. He knew that China had a long history of tea industry and had enjoyed a worldwide reputation in terms of the tea industry, which later declined and remained sluggish for a long time because of political corruption, backward productivity, wasted tea plantation and poor and unstable living, and thus the world's tea market had been taken up by India and Sri Lanka. Therefore, he decided to devote himself to the tea industry of China. He once visited Japan to learn modernized tea technologies. After he finished schooling, he started to run about for the rejuvenation of China's tea industry. He cooperated with his friends in establishing the Tea Export Inspection Institution, formulating the Plan of Rejuvenating China's tea industry, establishing the first experiment farm and the first tea research institute of China, as well as the country's first department of tea science at Fudan University. Also, he traveled to countries including India, Sri Lanka, Indonesia, Japan, the U.K. and Russia

in order to learn from the most advanced experience from these countries. He worked hard to explore the ambitious plans for rejuvenating China's tea industry. He left his footprints all over China for the tea industry. He has contributed a great deal to the tea, thus he is reputed as the "Sage of Tea in Contemporary China" by Lu Dingyi, one of the former state leaders of China.

20. What are the contents of the "Chinese Morals of Tea" advocated by Mr. Zhuang Wanfang, a famous tea science scholar?

Professor Zhuang Wanfang has put forth the proposition of "Chinese Morals of Tea" in March 1989, the contents of which are "Honesty, Beauty, Harmony and Respect". According to the interpretation of Mr. Zhuang, "Honesty" stands for "like a cup of clean and clear tea, one must promote being honest, clean-fingered and frugal and cultivate morality; toast to your guests with tea and drink tea in place of liquor". "Beauty" stands for "like a cup of clean and clear tea, one shall mainly taste and appreciate it, share the aromatic tea with others, smell the aroma of the tea together with others, and communicate with close friends over a cup of tea while enjoying the happiness of longevity". "Harmony" stands for "Like a cup of clean and clear tea, one should attach importance to morality and endow tea with a meaning of gift-giving, get along with others well in good faith and create good interpersonal

relationship". "Respect" stands for "like a cup of clean and clear tea, one is supposed to respect others and love people, take pleasure in helping others, and ensure that the tea ware is purified and the water is of good source".

The Processing of Tea

Chapter Information Board

21. How to classify tea in modern times?

There are multiple classification methods for teas. Currently, the prevailing method is classifying the teas into "basic teas" and "reprocessed teas". Basic teas can be classified into green tea, yellow tea, dark tea, white tea, cyan tea (Oolong tea) and black tea (which are in general called "six types of tea") according to the processing principles and quality features. The reprocessed teas mainly refer to scented teas and compressed teas, etc.

22. What are the procedures for processing green tea?

The primary processing of green tea includes four procedures, i.e., spreading, fixation, rolling, and drying. Fixation is the key step in the processing of green tea. The purpose of pan-frying is to utilize high temperature to deactivate polyphenol oxidase and thus prevents the oxidization of tea polyphenol substances, resulting in the special quality feature of green tea.

23. What are the quality features of green tea?

The quality feature of green teas is "clear infusion with green leaves". Amongst the teas produced in China, green teas are featured with the most categories. Based on the different methods of deactivating enzymes and drying, green teas can be classified into "fried green tea", "steamed green tea", "baked green tea" and

"sun-dried green tea". Currently, there are several thousands of brand names of green teas produced in China.

24. How to distinguish "fried green tea" from "baked green tea"?

The differences of "fried green tea" and "baked green tea" lie in the different methods used in the drying process in the elementary stage of processing. The frying process uses the method of frying in the whole drying process, featuring tools of boilers or frying machines. It needs to heat the boilers and evaporate partial moisture in the tea leaves through a heat transfer and thermal radiation in contact to achieve the purpose of drying. The baking process means using the method of baking in the whole drying process, featuring tools of baking cages or dryers. It uses coal pit or hot air generation furnace to produce hot air to evaporate moisture in the tea leaves to achieve the purpose of drying. The difference in the quality features of "fried green tea" and "baked green tea" is: fried green tea has a tight cord shape and a green and smooth color; its tea infusion is green and bright with highly fresh aromas and a chestnut aroma, offering a full-bodied and refreshing taste. The baked green tea, compared to fried green tea, has a rather loose cord, an obvious pekoe and an emerald green color with a refreshing aroma of orchid, offering a fresh taste.

25. What is the season when higher quality tea could be produced?

Green tea can be categorized into spring tea, summer tea and autumn tea according to the seasons of production. In general, spring tea has better quality. Thus, the consumers usually purchase green tea once in spring for the consumption of all year round.

The spring green tea has a full-bodied taste because, after a winter period of nutrition accumulation, the tea plants are sufficiently provided with nutrients, thus the tea leaves are teeming with the content of effective constituents.

Thanks to the improvement in tea varieties, agronomic and processing technologies, summer tea and autumn tea in some regions are of decent quality too.

26. Where is West Lake Longjing Tea produced? What are its quality features?

West Lake Longjing (Dragon Well) Tea is produced in the West Lake District of Hangzhou City, Zhejiang, China. The tea is characterized by four unique quality features, including green color, intense aroma, full-bodied taste, and gorgeous look. Moreover, the tea has a flat and sharpened shape, a smooth and even texture, and a light green and smooth color. The infusion of Longjing tea has a light and bright green color, fresh and brisk mellow and sweet

aftertaste, and fresh and lasting aroma. The steeped Longjing tea has tender leaves of flower-like shape. According to the regulation of the National Geographical Indication, Longjing tea produced only within the scope of 168 square kilometers of West Lake Production Area can be called "West Lake Longjing Tea".

27. What are the production sites of Longjing Tea?

According to the code *GB/T 18650-2008 Product of Geographical Indication-Longjing Tea*, the Longjing tea-producing sites can be categorized into West Lake Production Area, Qiantang Production Area, and Yuezhou Production Area. The administrative region governed by the West Lake District (West Lake Tourist Destination), Hangzhou City, belongs to West Lake Production Area. The administrative regions governed by counties including Xiaoshan, Binjiang, Yuhang, Fuyang, Lin'an, Tonglu, Jiande, and Chun'an in Hangzhou City belong to Qiantang Production Area. The administrative regions governed by counties including Shaoxing, Yuecheng, Xinchang, Shengzhou, and Zhuji in Shaoxing City and part of the townships and towns, including Shangyu, Pan'an, Dongyang, and Tiantai belong to Yuezhou Production Area.

28. Is "Anji Baicha" a type of white tea?

Anji Baicha is a kind of green tea produced in Anji County, Zhejiang Province. The tea is processed from the tender shoots of 'Baiye 1' tea leaves, an albino tea cultivar. It is a natural mutant sensitive to low temperatures. Typically, when the average temperatures in spring last between 19-22℃, the germinating buds and leaves lose chlorophylls and turn white. However, when the temperatures remain above 22℃, the tender shoots gradually turn green from white, which looks similar to the other green tea cultivars. Anji Baicha is picked, processed and made during a particular whitening period when the air temperature is relatively low. In fact, the processing procedure of Anji Baicha is the same as that of green tea. It has a straight and relatively flat look, like an orchid; it is green in color and slightly tippy; its bud leaves are like green sheaths embroidered with golden edges. Once it is brewed, it

gives off an intense aroma with a lasting fragrance and a fresh and brisk taste. In this sense, Anji Baicha is not white tea but green tea.

29. Is Jinyun Huangcha a type of yellow tea?

Jinyun Huangcha is exquisitely made from the new shoots of a special variety, 'Zhonghuang 2', which is a natural mutant. The processing procedure is similar to that of green tea. Thus, it is a type of green tea, with an amino acid content of above 6.5% and the total polyphenol content of 14.7%-21%. It has a fresh and brisk taste and a thick tea flavor. When brewed in a glass, it has a clear infusion and bright leaves, offering an extremely high value of appreciation. Tiantai Huangcha (made from the new shoots of 'Zhonghuang 1', also a natural mutant) is similar to Jinyun Huangcha. The traditional yellow tea, as one type of the Six Major Categories of Teas of China, is formed in the "heaping" procedure. It is characterized by yellow infusion and yellow tea leaves, such as "Junshan Silver Needle".

30. How is black tea processed?

Black tea is a type of tea, which is manufactured from fresh young shoots of tea plants through a series of steps, including withering, rolling (cutting), fermentation, and drying. Fermentation is the key processing step in black tea produetion. The essence of fermentation is the biological oxidization of tea polyphenols catalyzed by a number of enzymes, including polyphenol oxidase. Theaflavins and thearubigins are the main products of oxidization and polymerization of tea polyphenols (including EGCG, EGC, ECG, EC, etc.), which characterize the quality features of black tea. Compared to fresh shoots, the contents of total polyphenols in the made black tea decrease by 90%, whereas the kinds of aroma substances and their quantity significantly increase. Therefore, black tea has the features of "red infusion with red leaves", sweetness, and a full-bodied taste.

31. Why is the "red tea" in Chinese called "black tea" in English? What are the quality features of black tea?

Black tea is a type of fully fermented tea featuring the quality of "red infusion with red infused leaves". Thus it is called "red tea" in Chinese. Dry black teas of good quality often have black bloom color, except golden hue in the buds. Thus, it is called "black tea" in English because it looks black when it is dry. The quality features of black tea are a shiny black color of dry tea, a bright red infusion, and coppery-color steeped leaves. The black tea produced in China can be classified into Congou Black Tea, Souchong Black Tea, and CTC (Crush Tear Curl) Black Tea. Each of them has distinct features. Congou Black Tea is featured with a tight and slender shape, high aroma, and mellow taste. Souchong Black Tea is featured with fat and bold shape, and a slight smoky pine aroma.

The CTC Black Tea is featured with clearly graded shapes, including leaves, pieces, blades, and powder. All of them have a fresh and intense taste. In addition, turbidity frequently occurs when the infusion of black tea is cooled, a phenomenon called "cream down". The turbid matter is the complex reaction product of theanine, theaflavins, and thearubigins, etc. The phenomenon of "cream down" indicates a good quality of black tea.

32. What are the technological features of Oolong tea?

Oolong tea is a kind of semi-fermented tea processed from fresh shoots of a certain degree of maturity. Processing of Oolong tea includes steps of sun-drying or withering, rotating, air-cooling, heating, rolling, and drying. Air-cooling and rotating are unique techniques for making Oolong tea.

33. What are the quality features of Oolong tea?

Oolong tea is produced mainly in Fujian Province, Guangdong Province, and Taiwan Province in China. It can be further classified into Northern Fujian Oolong, Southern Fujian Oolong, Guangdong Oolong, and Taiwan Oolong; nonetheless, each has different quality features according to the places of production. The common features of traditional Oolong teas are sand green and black bloom dry leaves, high and obvious natural flowery aroma, golden yellow infusion, a full-bodied taste, and green infused leaves with red edges (so-called green leaves embroidered with red edges or jade plates with red edges). The red edge of green leaves occurs as the result of partial oxidation of polyphenols when the leaf edges crash and break during the rotating and air cooling steps.

34. What is the production area of "Wuyi Rock Tea"?

Wuyi Rock Tea is a type of Oolong tea made of young shoots from local varieties planted within the specified areas of Wuyi Mountain in unique natural conditions and is processed with traditional techniques. Wuyi Rock Tea carries a distinctively thick floral aroma (called as *Yanyun*).

Wuyi Rock Tea is produced in Wuyishan City, Fujian Province. There are plenty of rocks on the mountain. The lush vegetation provides abundant organic matter, and the weathering of rocks provides abundant mineral elements to tea plants. Most tea plants of this area grow on the slope sediments of rock mountains, and thus, the tea is called "Rock Tea".

The quality features of Wuyi Rock Tea are a strong and even cord, a sand green color with little white spots like that of frog skins, a lasting and intense aroma with a flowery fragrance, an orange infusion, a full-bodied taste, and a refreshing aftertaste, which is a unique flavor of Rock Tea. Moreover, it is characterized by thick and soft leaves, and the edges of which are red, and the middle areas are light green, thus presenting "green bellies with red edges".

35. Where are "Three Pits and Two Ravines"?

Wuyi Mountain is teemed with "Rock Tea". The mountain geography results in diverse micro-climatic and ecological conditions, where many tea plantations are located. Consumers prefer Wuyi rock teas produced in places called Three Pits and Two Ravines, which refer to Huiyuan Pit, Niulan Pit, Dakengkou Pit, Liuxiang Ravine, and Wuyuan Ravine. These places are the traditional authentic plantation areas, where the most famous Rock Tea is produced.

36. Is "Dahongpao" (Red Robe Tea) a black tea?

"Dahongpao" (Red Robe Tea) in Chinese has the character "red" in its name, but it is not a really black tea ("red tea" in Chinese). Dahongpao is one type of Wuyi Rock Tea, i.e., a type of Oolong tea amongst the Six Major Categories of Chinese Teas. Nowadays, the processing of Dahongpao still carries the traditional handcraft that includes five major procedures: withering, rotating, deactivating enzymes, rolling, and baking. These procedures can be further divided into 13 sub-procedures, i.e., withering, de-enzyming (rotating, flipping, resting), frying, rolling, second frying, second rolling, first baking, winnowing, cooling, picking, second baking (with adequate heat), wrapping-up, and additional heating. The whole process of rotating requires controlling the extent of leaf crashing and breaking. The traditional standard of making Rock

Tea requires that at the end of rotating step, 70% (area) of leaf in the middle remains green while another 30% leaf in the edges turns into red as the result of oxidation of polyphenols. Either too severe or too light oxidation will affect the quality of Dahongpao. Dahongpao has a twisted, strong and tight cord, with a color of bloom greenish brown, and an intense and lasting aroma, bringing a deep and remote feeling to drinkers. It has a full-bodied taste, a fresh and smooth aftertaste, an obvious flavor of Rock Tea, and also leaves a lasting aroma on the bottom of cups. Its infusion is dark orange and clear. The steeped leaves are soft and bright, even and orderly with distinct red edges.

37. What are the processing techniques and quality features of white tea?

As one of the six major categories of tea of China, White tea is a kind of slightly fermented tea. The traditional processing of white tea has two steps, withering and drying. White teas can be further divided into "white tip silver needle", "white peony", "Gongmei" and "Shoumei" according to the tenderness of young shoots. Due to the unique manufacturing technique that lacks frying or rolling in processing, the "white tip silver needle" tea has a fairly tippy look, a fresh aroma, yellowish green and clear infusion, a full-bodied taste, and a sweet aftertaste. It is mainly produced in Fuding, Zhenghe, Jianyang, and Songxi in Fujian Province. White tea is characterized by curing efficacies including cooling and refreshing, bringing down a fever, decreasing internal heat and driving away summer heat and also a quiet, simple but elegant character.

38. What are the processing techniques and quality features of yellow tea?

Yellow tea is made through the heaping and smothering procedure, the so-called yellowing step, by covering a pile of warm and wet de-activated young shoots to facilitate the change of leaf color from green to yellow. The unique quality feature of this tea is "yellow infusion with yellow leaves", and it is a slightly fermented tea.

39. What are the processing techniques and quality features of dark tea?

Post-fermentation of made teas at piling is a unique and key procedure in the processing of dark tea that contributes to the formation of its color, aroma, and taste. The fresh leaves for dark tea are relatively coarse and aged. After piling and fermenting for a rather long time during the processing, the color of tea leaves turns into bloom dark and auburnish black; thus, it is called "dark tea". Quality dark tea features an aged and pure aroma, a mellow and smooth taste, a bright color of the brew, ideally in orange, orange-red or red, and a bright brownish-black color of steeped tea leaves.

40. What is Pu'er tea?

Pu'er Tea is a special tea made only in Yunnan Province. It is made from the sun-dried green teas from local large-leaf varieties, with special processing technology within the protection scope of geographical indications. It has unique quality characteristics. According to its processing technology and quality characteristics, Pu'er tea can be divided into two types: raw tea and mature tea. According to its appearance, Pu'er Tea can be divided into loose Pu'er tea (post-fermented) and compressed Pu'er tea (post-fermented or unfermented).

41. What is the post-fermentation of Pu'er tea processing?

Pu'er tea is made of sun-dried green tea of local large-leaf varieties in Yunnan Province after a specific process called post-

fermentation. In the post-fermentation, the sun-dried green tea is piled up with the addition of some water and fermented naturally for some time under the heat and humid conditions. With the aid of local special microorganisms and the excreted enzymes, the quality ingredients within green tea go through a series of oxidation and transformation to produce the unique features of ripe Pu'er tea.

42. What are the "golden flowers" in Fu brick tea?

The processing of Fu brick tea features a unique solid fermentation aided by the fungi *Eurotium cristatum*. *E. cristatum* is inoculated artificially to the green tea and multiplied under the favorable temperature and humidity, thus forming a yellow cleistothecium, i.e., the "golden flower". The fungal fermentation plays a significant role in forming the unique taste of the Fu brick tea and its health benefits.

43. Is the tea rich in pubecence good or not?

The key factor that determines the quantity of tea pubescence, i.e., trichomes, is the variety of tea plants. Some varieties have characteristically many pubescences, yet some barely have any pubescence. The degree of tenderness also affects the quantity of tea pubescence. Generally, tender young shoots have more pubescences than mature shoots. Therefore, for the same variety of tea, the one with more pubescences generally serves as better raw material. The processing techniques also affect the quantity of tea pubescence. For example, one purpose of pan frying of Longjing tea is to get rid of its pubescence. The frequent stir-frying and churning for more times also make teas have fewer pubescences. The tea pubescence contains substances including amino acids, which are, to a certain degree, conducive to the qualities and nutrient content of made teas. Therefore, the quantity of tea pubescence is mainly decided by the variety of tea, and is also affected by tenderness and processing techniques. So tea cannot be judged good or bad only based on its quantity of pubescence. Because some good teas have no pubescence, while most teas, with more tea pubescences can be of good quality as well.

44. Are there microbes in dried tea?

Microbes exist everywhere in daily life. Accordingly, microbes exist in tea and they play an important role in the tea production process. For example, post-fermentation, in which microbes are involved, is an important part of dark tea processing. Studies show that a variety of microbes, including Aspergillums, black mold, and yeast are isolated and identified in the dark tea. The activities of these microbes promote the formation of the unique qualities of dark tea.

45. What is Matcha?

Matcha refers to the powder-like tea product which uses fresh tea leaves cultivated under covers. After de-activation by steam (or

hot wind), the tea leaves are dried and then processed through the grinding technique.

46. What are the quality features of Matcha?

Matcha is not only characterized by a fresh and green infusion and a unique flavor of sea sedge, but also features high contents of amino acids, chlorophylls, and other special elements of nutritional and healthcare values. It has low bitterness and a full-bodied taste. Matcha is widely used as an additive in the modern food industry.

47. How is instant tea powder processed?

Instant tea powder is a solid tea beverage that can quickly dissolve in the water. It uses finished tea, semi-finished tea, by-products of tea or fresh tea leaves and other herbaceous plants and grains as raw materials. Through production techniques including extraction, filtration, purification, concentration and drying, it is then processed into a new drink which is crystal-like, powder-like or tablet-like, and dissolvable, leaving no tea grounds. The instant tea powder product can be categorized into two types, instant pure tea and instant seasoning tea. It has many features, including being easy to brew and portable, dissolvable, leaving no tea grounds, little pesticide residues, easy to season, and easy to blend with other food. Thus, it is increasingly more promoted in the tea market.

The Drinking of Tea

Chapter Information Board

48. What are the fundamental elements for brewing a good cup (pot) of tea?

The brewing of a good cup of tea will be affected by a variety of factors. Regardless of the quality of the tea, different tea brewers have their brewing habits. The experienced tea brewers know the features of the tea they brew. Thus they can choose suitable brewing methods based on the nature of tea, which usually include controlling the leaching efficiency of the flavoring ingredients to accentuate the enjoyable elements of the tea's features, and to reduce the dissatisfaction, if any, caused by the defects in the tea's quality. The fundamental elements of brewing tea can be concluded in the following six aspects:

① Quality of tea: tea quality is the prerequisite for a better taste of tea infusion;

② Choice of water: the mineral elements or other substances contained in the water can affect the brewing;

③ Teaware: the shapes and materials of teaware affect brewing;

④ Brewing time: the different lengths of brewing time can lead to different infusion;

⑤ Brewing water temperature: under different temperatures, the flavoring and aromatic substances in the tea have different leaching efficiency and volatilization rate, thus resulting in different flavor;

⑥ The ratio of tea leaves to water: different ratios can result in different brewing results.

49. How to choose the time for brewing tea?

The time used to brew tea has an evident effect on the taste of infusion. The longer the brewing time is, the higher their contents are in the infusion. Also, the concentration of taste is positively correlated with the content of compounds leached into the infusion; thus, the longer the brewing lasts, the more concentrated the infusion is; the shorter the brewing lasts, the less concentrated the infusion is. People differ in their preference of the taste concentration. Some people like to drink strong tea, and yet some people like to drink mild tea. But the concentration of infusion, which most people can accept, is variable within a certain range. Therefore, you can choose how to brew a cup of tea according to your own flavors.

50. What is the requirement for water temperature for brewing tea?

Generally, the solubility of the content of compounds leached into the infusion in the water is positively correlated with the temperature of the brewing water. The higher the water temperature is, the more the contents are leached, and the more concentrated is the infusion. Conversely, the lower the water temperature is, the less concentrated the infusion is. However, the requirements of temperature of brewing water are different for different types of tea. When brewing premium green teas of tender buds and leaves, the temperature of brewing water is better not too high, ideally between 75-85°C, because only with water at a relatively lower temperature, the infusion can become light green and the taste can become refreshing. By contrast, brewing with too hot water increases the leaching of tea polyphenols, likely to make astringent taste and damage vitamin C. The choice of water temperature varies with the consumers' geographical regions, ages, genders, habits, etc. Generally speaking, water at high temperature, for example boiling water, is used for Oolong tea, dark tea,

white tea and scented tea. For black tea and yellow tea, the temperature depends on how tender the leaves are: lower temperature like 75-85°C for tender leaves and higher temperature like 85-90°C for those that are relatively less tender. In any cases, the above mentioned temperature of water is attained by cooling down boiling water.

51. What is the appropriate tea-to-watere ratio in brewing?

Generally to say, placing more tea means a stronger taste. An appropriate tea-to-water ratio may rest on the tea variety, the size of teaware and the personal taste. For example, the ratio can be approximately 1:50-1:75 in brewing premium green tea or black tea, i.e. 3g of dried tea requires 150-200mL of water between 75-85°C, but in brewing Pu'er tea or Oolong tea, the amount of tea should be twice as that of black tea and green tea for the same size of water. People of ethnic minorities, who have a penchant for "brick tea" that features highly concentrated infusion and efficacies of decomposing fat and facilitating digestion, like to follow a ratio between 1:30 and 1:40 by applying 1,500-2,000mL of water to 50g of brick tea.

52. Why should we separate infusion from tea?

It is recommended to separate infusion from tea in brewing fundamentally with following reasons. First, as different chemical components of tea may offer different leaching rates, the separation will keep the flavor of each brew approximately the same by the control of steeping time. Second, since tea varieties differ in an acceptable number of brews, the action will give us fully a sensory joy from each brew. Third, it will make tea better-tasted and more

nutritious, which can be more instrumental to maintaining the color, taste and nutritions of the infusion.

53. Shall we discard the first brew of tea?

Some people believe that the first brew of tea is not clean, so they always discard it when brewing tea, believing that through this, they can clean the dust in the teaware and rid the pesticide residue. However, this practice is mistaken. Currently, the production of tea can be mechanized, continuous, and clean; in this case, the tea produced are clean and sanitary. Notably, the tea sold in the market must meet the China's Food Safety Standard. As long as the consumers buy qualified tea products, the sanitation indicators of the tea are trustworthy, and consumers can drink the tea with reassurance. What is more important is that most amino acids such as theanine, vitamin C, and some other nutritional substance are already leached into the first brew, so discarding the first brew of tea leads to the loss of these most nutritional substances. Therefore, the first brew of tea should not be discarded.

54. Do the different qualities of water influence the quality of the brew of green tea?

Different kinds of brewing water bring along different flavors of green teas. Bottled drinking water such as purified water (distilled water), natural spring water, natural mineral water, and tap water or source water are frequently used to brew a cup of tea in daily life. Normal brewing with purified water (distilled water) renders the pure and original flavor of green tea. Natural water (spring water) can better the taste and aroma of the tea to a certain extent. Mineral water influences the flavor of green tea, mostly in a negative way. The quality of tap water varies greatly for its sources. Generally speaking, the flavor of tea brewed by tap water in large cities is poor. Alkaline water is not suitable for brewing green tea. Purified water (distilled water) is a normal choice, and natural water (spring water) of low mineralization, low hardness, and low alkalinity is a good choice for fastidious drinkers.

55. Sometimes, white precipitates may appear in brewed green tea. Is there a problem with the tea?

White precipitates occur mainly due to the property of the brewing water. Tea leaves contain a high content of oxalic acid. Therefore when the water is too hard, having a high concentration of calcium and magnesium ions, large amounts of insoluble white

precipitates, for example, calcium oxalate and magnesium oxalate, are produced. It is not an indication of the tea itself.

56. How to brew green tea in glasses?

When brewing green tea in glasses, the consumers can choose different methods, such as dropping tea at last, dropping tea secondly or dropping tea firstly. Brewing with dropping tea last means pouring water in the glass first, and then dropping tea into the water. In this way, tea gradually stretch on the surface of the water and slowly descend. The advantage of this brewing method is that the hairs on tea are maximally preserved. Thus this method is suitable for brewing tender teas with dense hairs. Dropping tea secondly means pouring one-third of water into the glass, then dropping tea into the water, and finally filling the glass with water. This brewing method is suitable for tea which is neither too tender nor difficult to descend. Dropping tea firstly means dropping the tea into the glass first, and then pouring the water. This is suitable for the teas made of large buds or leaves which are difficult to sink to the bottom.

57. How to brew black tea?

Black tea can be brewed in many different ways. Bagged black tea and instant black tea normally are brewed in a cup. Broken black tea (CTC), black tea tablets and black tea powder are

normally brewed in a pot, which is conducive for separation of infusion from the tea dust. Congou black tea and Souchong black tea can be brewed in a bowl with a lid or in a pot.

Take Kungfu black tea as an example. It is usually brewed in a lidded bowl. For each 3 grams of tea, it takes 150 mL of water at 85°C. Allow 45 seconds for the first brew, 20 seconds for the second brew, and 40 seconds for the third brew. Adding milk and some sugar into black tea brew makes the different flavors and is liked by some foreigners. Some people are fond of mixing infusions with ice, honey, and fresh lemon in utensils to make iced tea.

58. How to brew Chaoshan Congou tea?

Congou tea, as an extremely elaborative tea-drinking habit, prevails in southern Fujian Province and Chaoshan Area of Guangdong Province.

A set of antique tea wares called "Four Treasures for Brewing Tea" is a prerequisite to enjoy Chaoshan Congou tea. The first treasure is Yushu Kettle, an ochre-brown platode water kettle with a capacity of around 200 mL. The second treasure is Chaoshan Stove, which is used to boil water. The third treasure is Mengchen Pot, a type of ceramic teapot as big as a goose egg with a capacity of around 50 mL (50 g water). The fourth treasure is a tiny Ruochen Cup, half of a ping-pong ball in size, holding 10-20 mL of infusion. Taking "Fenghuang Dancong" as an example, 5 g of "Fenghuang

Dancong" is brewed with 100 mL boiled water at 100°C. It allows 30 seconds for the first brew, 20 seconds for the second brew, 30 seconds for the third brew, and an extra 10-15 seconds for each following brew. This tea can be brewed 5-8 times.

59. How to brew yellow tea ?

The yellow tea is normally brewed in a lidded bowl or in a glass or a pot. The method and water temperature for brewing yellow teas (bud and small leaf types) are similar to those for brewing green teas. Tender yellow teas with good shape can be brewed in a glass or a lidded bowl. For example, three grams of Mogan yellow-bud tea is brewed with 150 mL boiled water at 80°C. The first brew takes 80 seconds, the second brew is 50 seconds, and the third brew spends 60 seconds. The yellow tea of big leaves can be brewed in boiled water at a temperature over 95°C.

60. How to brew white tea?

White teas can be brewed in lidded bowls by dropping the tea first. The water temperature should be high (better over 95°C), especially when brewing "Baihao Yinzhen"(White Tip Silver Needle), and a longer infusion time than for other types of teas should be used. Medium and large-sized lidded bowls are chosen to brew loose white teas like "Baimudan"(White Peony) and "Shoumei". White porcelain cups are recommended to sip for appreciating the beauty of the infusion color. Take White Peony (Baimudan) as an example, each 5 grams of tea is brewed with 150 mL boiled water at 90°C, which allows 60 seconds for the first brew, 30 seconds for the second brew, 40 seconds for the third brew, 60 seconds for the fourth brew, and 80 seconds for the fifth brew. Brewing aged white tea in a large-sized pot makes the flavor even better.

61. How to brew dark tea?

The water temperature should be high while brewing dark tea. Normally large-sized boccaro teapots are selected to brew dark teas for their good retention of high temperature. Taking compressed Pu'er tea as an example, each five grams of dark teas is brewed with 100 mL boiled water at 90°C. The first brew takes 20 seconds, the second brew takes 10 seconds, and the third brew takes 15 seconds. Each of the following brews thereafter takes an extra 5-10 seconds, and dark teas can be brewed more than 7 times.

62. Why is tea astringent?

The polyphenols in tea contain free hydroxyls, which can

combine with the proteins in the oral mucosal epithelial tissues. The complexes are solidified into an impermeable film, which produces a taste of astringency. If the polyphenols contain many hydroxyls, a thick impervious film is formed to give a strong astringent taste as eating raw persimmons. If the polyphenols contain limited hydroxyls, the formed impervious film is thin and unstable and gradually dissociates, producing an astringent taste at the first and sweet aftertaste.

63. Why do some teas have grass odor?

The main ingredient that produces grass odor is geraniol, a terpene alcohol. It exists in fresh tea leaves with a boiling point of 156°C. The white tea processing is simple and mainly involves withering and baking, and thus geraniol is not completely evaporated. This is especially true for new white teas (within the first year after production), which usually carry a grass odor. Green tea without sufficient inactivation of enzymes would also have a type of grass odor.

64. Can children drink tea?

Yes, but weak tea is suggested. Since the caffeine in tea can be refreshing to human brain, an excessively strong tea may affect children's nervous system and cause over-excitement. It is therefore recommended for children not to have strong tea.

65. Is overnight tea drinkable?

Overnight tea can be taken in as long as it has not spoiled. Yet in summer, tea brews deteriorate easily to produce an unpleasant smell due to the hot weather. Sometimes tea brewed in the morning is not drinkable even in the afternoon. Many foods contain nitrates. These matters could combine with secondary amines to form nitrosamines inside the human body under certain conditions. Nitrosamines are carcinogenic during the process of metabolism in animals. As tea contain a certain amount of protein, some people surmise that the overnight tea would contain nitrites and thus is carcinogenic. In fact, a very little amount of nitrites is formed in the overnight tea, even if there is any. Moreover, nitrite per se is not carcinogenic and becomes a bit toxic only after forming nitrosamines by reacting with secondary amines under certain conditions. Moreover, tea contain abundant polyphenols and vitamin C, which hinders the formation of nitrosamines. Thus you can drink the previous night's tea if it does not deteriorate.

Of course, we do not encourage people to drink such previous-night tea. Any beverage tastes best when it is fresh. The same is true for tea. The newly brewed tea is fragrant, more nutritious, and contains less living contaminants. Thus, it is better to drink the newly brewed tea.

66. Can we swallow the medicine with tea?

Tea contain caffeine, theobromine and tea polyphenols, which might have chemical reactions with certain compounds in the medicine. This may influence the efficacy of the medicine, so it is not advisable to swallow certain kinds of medicine with tea or drink tea immediately after taking medicines. The depressant and hypnotic effects of Miltown, barbital, diazepam, among other inhibitors of the central nervous system, might be offset by caffeine and theophylline that can stimulate the central nervous system.

Caffeine in tea can contradict adenosine, consequently perils the efficacy of dipyridamole when it is taken together with tea by patients suffering from cardiovascular diseases or nephritis. The efficacy of chalybeate deteriorates if anemia patients have this medicine with tea because tea polyphenols can precipitate chalybeate. In addition, chlorpromazine, aminopyrine, pantopon, berberine, digitalis, biofermin, multienzyme tablets, pepsin, ferrous sulfate and tetracycline, and other antibiotics medicines can combine with tea polyphenols and form insoluble precipitates. This will affect the absorption of the medicines. In order to give full play to the performance of the medicines and avoid adverse consequences, one should not have medicine, no matter the Chinese medicine or the western medicine, with tea or drink tea right after taking medicine unless otherwise instructed by the doctor.

67. Why has tea become a necessity for minority ethnic groups in northwest China?

This is related to their living habits and geographical environments. There are three main reasons. First, they eat a diet heavy on dairy products and meat, which are fatty and difficult to digest. Some ingredients of tea help break down fats and promote digestion. Second, the altitude is high in northwest China. The thin air and dry climate cause a faster water loss in the human body and call for adequate water intake. Apart from providing water to the human body, drinking tea can stimulate saliva secretion with its polyphenols, which increase appetite and quench thirst. Third, people living in highlands suffer from a shortage of vegetables and fruits and are therefore susceptible to vitamin deficiencies. Tea contains various vitamins, which is acquired by people who drink tea.

68. How can we make dishes with tea?

There are usually four ways of cooking with tea: ①tea dish, mix fresh tea with dishes while grilling or frying; ②tea soup, stew or simmer dishes in the tea soup; ③tea powder: grind the tea into powder and sprinkle it into the dish or snack; ④tea fumigation, fumigate food with the aroma of tea.

69. Is the tea infusion acidic or alkaline?

This depends on the relative concentrations of free hydrogen ions and free hydroxide ions in the tea infusion. Main acidic substances in the tea are various carboxylic acids (such as citric acid, fatty acids), certain amino acids, vitamin C, theaflavins, and thearubigins among others; alkaline substances are mainly caffeine and some aroma compounds. The pH value of the  infusion has a certain correlation with the processing of tea, the quality of made tea, and the brewing of tea. More often, tea infusions are acidic, and there is no alkaline one.

The Benefits of Tea

Chapter Information Board

70. What are the chemical components of tea?

Over 700 different chemical compounds have been identified in tea. Fresh tea leaves harvested from tea plants contain 75%-78% moisture, and the rest 22%-25% dry masses consist of organic and inorganic components. Organic components mainly include the following (dry weight): 20%-30% of protein, 20%-25% of saccharide, 10%-25% of tea polyphenols, about 8% of lipids, 3%-5% of alkaloids, 2%-7% of free amino acids, about 3% of organic acids, about 1% of pigments, 0.6%-1.0% of vitamins, and 0.005%-0.03% of aromatic substances.

71. What are the effects of tea?

Research has shown that tea has 23 main effects: ①quenching thirst; ②relieving fatigue; ③relieving tension; ④strengthening heart; ⑤diuretic; ⑥preventing and lowering high blood pressure; ⑦preventing and lowering hyperglycaemia; ⑧protecting the liver; ⑨facilitating bowel movement; ⑩relieving greasy taste; ⑪combining to detoxify heavy metals (such as lead, arsenic); ⑫antibacterial activity; ⑬diminishing inflammation; ⑭antiviral; ⑮preventing gastrointestinal cancer, breast cancer, prostate cancer and lung cancer; ⑯antioxidant; ⑰anti-radiation; ⑱preventing dental caries; ⑲relieving hangover; ⑳preventing ophthalmic diseases; ㉑preventing calculus; ㉒preventing atherosclerosis; and ㉓preventing neurodegeneration.

72. What are the effects of tea polyphenols?

The main effects of tea polyphenols include: ①effectively enhancing functions of blood capillary, toughness of microvascular walls; ②facilitating the absorption of vitamin C and prevent scorbutus; ③detoxifying harmful metals to certain extent (such as reducing toxic hexavalent chromium ions to non-toxic ions; ④inhibiting atherosclerosis, and reducing the incidence of hypertension and coronary heart disease; ⑤having significant antibacterial and bactericidal effect; ⑥relieving hyperthyreosis; ⑦relieving anti-radiation damage and increasing the total number of white blood cells; ⑧inhibiting mutation and cancer cell generation; ⑨preventing peroxidation of intracellular lipid and significantly inhibit the formation of free radicals; ⑩having anti-coagulation and stasis removing effect, lowering blood lipids, preventing thrombosis from forming, and helping to lose weight; ⑪preventing neurodegenerative disorders and Alzheimer's disease; ⑫preventing ultraviolet rays' irradiation damage to the skin; and ⑬enhancing human immunity.

73. What are the effects of alkaloids in tea on the human body?

There are two main types of alkaloids in tea, namely caffeine and theobromine. Two alkaloids are methylpurine compounds,

which are important physiologically active substances and distinctive chemical substances in tea. They can significantly excite the central nervous system and have a diuretic effect. They are all capable of exciting the central nervous system. The alkaloid with the highest content in tea is caffeine, followed by theobromine.

74. What are the beneficial effects of theanine in tea on human body?

Theanine accounts for about half of the total free amino acids in tea, and is an important component of tea, especially closely related to the quality of green tea. Theanine has the following effects: enhancement of memory and learning ability; prevention of Parkinson's disease, Alzheimer's disease, and conduction nervous disorder; protection of the liver; reduction of blood pressure, tranquilization of the mind, and improvement of sleep; enhancement of immune function and delay aging.

75. What are the effects of γ-aminobutyric acid in tea on human body?

γ-aminobutyric acid is a non-protein amino acid in plants and animals. Generally, the content of γ-aminobutyric acid in tea is very low, but it can be significantly increased by special processing techniques.

γ-aminobutyric acid has significant effects on human health, such as lowering blood pressure, improving brain cell metabolism, and enhancing memory. It is reported that γ-aminobutyric acid can improve vision, lower cholesterol, regulate hormone secretion, relieve ammonia toxicity, protect the liver, and enhance liver function.

76. Does drinking tea improve dental health?

Tea has fluorine content, which can be utilized for improving dental health by drinking tea. Therefore, drinking tea in moderation is a reliable and effective method to prevent caries. Children under the age of 6 are in the period of permanent tooth germ formation. During this period, fluoride intake is beneficial for preventing caries. To this end, children can rinse their mouths with tea. The bactericidal effect of phenolics in tea can also improve the health of teeth.

77. Does tea have an antioxidant effect?

Tea has an excellent antioxidant activity, which is one of the critical features of its health benefits. The antioxidant effect of tea is closely related to its function of scavenging free radicals. There are many components in tea that can scavenge free radicals. The most important one among them is tea polyphenols. The three major mechanisms of tea polyphenols' antioxidant function are to inhibit the production of free radicals, to directly scavenge free radicals, and to activate the free radical scavenging system of the organism itself.

78. Is drinking tea helpful for cancer prevention?

Tea has a certain effect on cancer prevention. A large number of animals and population intervention studies have shown that regular and continuous drinking a certain amount of green tea can effectively prevent certain cancers. This is especially true for the prevention of digestive tract cancer, breast cancer, and prostate cancer. Its preventive effect on liver cancer is relatively good. Take digestive tract cancer as an example. Active ingredients from tea drinking widely are distributed in the digestive tract and metabolic organs, thereby inhibiting the absorption of carcinogens and promoting the apoptosis of tumor cells. The anti-cancer effect of tea is mainly ascribed to tea polyphenols, which can prevent cancer by inhibiting carcinogens (such as nitroso compounds), scavenging free radicals, blocking or inhibiting the growth of tumor cells, and promoting apoptosis of tumor cells. At present, a large number of epidemiological survey data show a better effect of tea in preventing cancer, especially in females. According to the study, the effect is counteracted somewhat by some bad habits such as smoking in males. In addition, tea is also rich in vitamin C and vitamin E, which can also help to prevent cancer.

79. Is drinking tea helpful for losing weight?

Population survey and intervention studies found that tea is good for lipid-lowering, but it does not have an obvious effect on losing weight. Drinking tea regularly can significantly improve the property of blood lipid and reduce cholesterol and triglyceride levels in the blood. In China, there are ancient records about tea's effect on losing weight, such as "good for losing weight", "dissolving oil", "frequent and long-lasting tea drinking making thinner", and so on. The lipid-lowering effect of tea is due to the comprehensive functions of various active ingredients (tea polyphenols, caffeine, vitamins, amino acids, etc.). The reason for obesity is that the absorption and synthesis of fat surpass its catabolism in fat cells. Tea polyphenols can decrease the absorption of fat of foods, inhibit the activity of enzymes (glycosidase, lipase, amylase, etc.) in the digestive tract, promote intestinal peristalsis

and lipid excretion, increase energy expenditure in the mitochondrion and finally help to lose weight.

80. Does tea have a whitening effect?

Polyphenols in tea can protect the skin from the damage of ultraviolet radiation, which wins them the title of "ultraviolet filter". Research shows that tea polyphenols protect skin from ultraviolet-induced damage more significantly than vitamin E. Tea polyphenols also inhibit the activity of tyrosinase and reduce the production of melanin by slowing the metabolism of melanocytes, thus whitening the skin.

81. Does drinking tea help prevent atherosclerosis?

The main cause of atherosclerosis is the oxidation of low-density lipoprotein, which strongly inhibits the movement of phagocytes, increases the retention of macrophages in the arterial wall, induces endothelial cell dysfunction, promotes the progression of fatty streaks into atherosclerotic lesions, and thus causes atherosclerosis. Tea polyphenols, the main functional ingredient in tea, can regulate the function of cells in the arterial wall, impede the absorption of cholesterol, inhibit the oxidation of low-density lipoprotein, and increase the proportion of high-density lipoprotein. Thus, drinking tea can prevent atherosclerosis.

82. Does drinking tea help prevent radiation damage?

Tea drinking can prevent radiation damage. In 1962, Soviet scholars injected mice with tea extract. After gamma-ray irradiation, most mice injected with tea extract survived and those without injection died. Moreover, these scholars also found that the extract of tea can protect hematopoietic function. Among the victims of the atomic bomb of Hiroshima in Japan during World War II, those who drank tea for a long time were less affected by radiation and had higher survival rates. Around 1973, researchers in China proved that tea extract could prevent the decline of white blood cells caused by radiation damage, which is conducive to the normalization of hematopoietic function. The main compounds with anti-radiation effects include tea polyphenols, lipopolysaccharide, vitamin C, vitamin E, cystine, cysteine, vitamin B, etc. Although the mechanism preventing

radiation damage has yet to be further studied, the good effects of tea drinking are obviously observed in many animal studies and clinical tests.

83. Does drinking white tea cure a cold?

White tea can help relieve inflammation or internal heat, but it cannot replace medicines. This is because: ①White tea does not go through the process of heating and rolling. The withering for an extended period under low temperature also causes different compositions from other teas. These characters enable white tea to be anti-inflammatory and heat-relieving; ②White tea contains a high level of flavonoids, which are natural antioxidants and can improve immunity and protect cardiovascular function. Generally, white tea stored for a long time has a higher content of flavone. Drinking white tea at the initial stage of a cold is more helpful; ③The function of white tea is mainly in improving immunity and prevention, yet it cannot become a substitute for medicine. If one suffers from severe inflammation and cold, it is better to consult with a physician. After all, body conditions and the situations of a cold vary greatly.

84. Is it good to gargle with tea after a meal?

After meals, various food residues are often left between the oral spaces, and the carcinogens such as protein toxins and nitrites can be produced by the action of biological enzymes and bacteria in the oral cavity. These substances can enter the digestive tract by drinking water, eating, swallowing and other oral movements, which are detrimental to the human body. If one gargles with tea after a meal, fluorine and tea polyphenols can inhibit the growth of bacteria between the interdental spaces. In the meantime, tea has an anti-inflammatory effect and can inhibit the proliferation of colibacillus and staphylococcus. Tea can also make the carnivorous fibers embedded in the teeth shrink and leave the tooth gap. Therefore, it is beneficial to using tea to gargle after a meal, especially after eating greasy food.

85. Does drinking tea have an anti-aging effect?

According to the free radical theory of aging, aging is the result of the loss of balance between free radical production and removal. Therefore, reducing the generation of free radicals or removing existing free radicals can effectively slow down the aging and wrinkling of the skin. Tea polyphenols are natural antioxidants with strong antioxidant capacity. Their ability to scavenge free radicals exceeds that of vitamin C and vitamin E. Therefore, tea can effectively prevent and alleviate skin aging and has a cosmetic effect.

86. What is the correct opinion on the efficacy of the tea?

Although tea has many healthy and medical effects, it is only a

beverage rather than a medicine. For people with obvious organ damage or who have been diagnosed with certain diseases, if they have any physical discomfort, going to the hospital to see a doctor in time is definitely the most important step to do. People should follow the doctor's advice as to whether they can drink tea during treatment or medication. Maintaining good health is inseparable from good living habits and balanced diets. Regular tea drinking helps to maintain good health and reduce the incidence of physical discomfort.

87. What is the foam on the surface of the tea at brewing?

There is often some foam floating in the infusion after brewing tea. Such foam is induced by tea saponin, which has high water-solubility and strong foaming capability. Saponin is quickly leached, especially when brewing water of high temperature is flushing to turn over tea, causing the formation of rich foam on the surface of the infusion. Tea saponin with anti-inflammatory and analgesic effects is one of the important active ingredients in tea.

The Storage of Tea

Chapter Information Board

88. What are the major factors influencing the quality change of tea during its storage?

During the storage of tea, substances like tea polyphenols, amino acids, lipids, vitamin C, and chlorophyll are very likely to be oxidized and degraded, thus causing changes in sensory qualities, including the color, aroma, and taste of tea. These changes are mainly subject to factors such as water content in tea, ambient temperature, humidity, abundance of oxygen and light intensity. Higher water content in tea, higher ambient humidity and temperature lead to a greater loss of quality. Abundant oxygen and light exposure can accelerate the oxidation of various quality substances in tea and lead to a faster loss of quality. In addition, due to its high absorptivity, tea is very likely to absorb peculiar smells. So odors must be avoided in the storage environment of tea.

89. How to store tea at home?

There are three ways to store tea at home.

① Storing tea in a container: Use an iron box, glass bottle or rice jar that can be sealed and of appropriate size to store tea. Place some lime or silica-gel as desiccant at the bottom, wrap tea with craft paper or into small packs and place them on top of the desiccant, which should be replaced every once in a while.

② Storing in small sachets: Store tea in the plastic compound bag with an effective sealing option, e.g., aluminum foil bag, add a small amount of desiccant and seal the sachet before putting them into an iron caddy or paper caddy.

③ Storing in a refrigerator: Taking advantage of the low temperature to maintain a stable quality of tea can be an economical and effective storage method. However, when using household refrigerators, given that tea can absorb smells and moisture quite easily, it is particularly important to seal the tea in packages so as to prevent the tea from absorbing smells from other foods in the refrigerator.

90. Is tea drinkable when it gets damp?

It depends on the time of exposure to a damp condition and how damp it becomes. If the period is short and the tea is not too damp and has not deteriorated, it would be helpful to immediately dry the tea (by measures such as drying and frying) to remove excess water. The tea is still drinkable, but the quality may be compromised, for example, the brew may become yellow and less fragrant. If the period is too long or when the tea becomes too damp, the tea may be deteriorated or even mildewed. Then the tea becomes undrinkable and should be thrown away.

91. Do different teas require different storage environments?

Different teas need to be stored in different environments according to their own characteristics. It is recommended to store non-fermented green tea, light-fermented South Fujian Oolong tea, Taiwan Oolong tea under refrigeration. An appropriate amount of tea should be taken out from storage, and when it reaches room temperature, it can be brewed to drink. Like black tea, yellow tea, Guangdong Oolong tea, and North Fujian Oolong are heavily fermented or baked. It is recommended to seal them in an environment at room temperature and protect them from the light. As for white tea, dark tea and other teas that need to be post-matured, they should be wrapped in cotton paper and stored at room temperature and away from light. Moreover, the ambient humidity should not be too high; otherwise, the tea leaves are easy to mildew.

92. Why is the tea quality easily affected?

The color, aroma, and taste of teas are easily affected and even deteriorated by the conditions of the external environment in the process of storage, transportation, and consumption. There are three main reasons. First, the loose and porous structure of teas endows them with strong adsorption ability. Second, the rich polyphenols in the tea are easily oxidized. Third, the main substances of the color, aroma and odor in the tea are easily oxidized and degraded.

93. How should premium green teas be stored?

Premium green teas are very susceptible to quality loss during their storage, and they need very particular storage conditions. The storage, transportation and consumption of premium green teas involve three major processes: large-scale storage, retail tea storage, and household storage.

① Warehouse for large-scale storage. Tea in large quantities is usually stored in dedicated cold storage units or warehouses that are dry and odorless. With a temperature between 2-8°C and a relative humidity below 60%, the quality of tea can be maintained stable for a storage period of 8-10 months within the warehouse.

② Retail tea storage. Retail tea products in small packages should be stored in compound bags with excellent sealing performance, e.g., aluminum foil bags. Fresh-keeping techniques should be taken. These can be desiccant or modified-atmosphere technologies such as oxygen removal or nitrogen filling. When possible, retail tea can be stored in small refrigerators, which has better storage effects.

③ Household storage of premium tea. These teas can be stored in refrigerators or freezers and should be closely sealed and kept away from other odorous foods.

94. How to judge whether the tea is deteriorated or not?

The deterioration of tea means a significant change in quality, which causes obvious flavor abnormalities. Whether there is a deterioration or not is usually judged by the color, aroma and taste of the tea.

① Observing the color: whether there is a significant change in the color of tea. Green tea turns from emerald green to yellow and dark. Black tea and other teas with a higher fermentation degree are covered with a gray layer.

② Smelling the aroma: the deteriorated tea becomes much less aromatic and lacks authentic smell. The aroma becomes light and difficult to be distinguished. Sometimes there are even unpleasant smells such as stale or rancid odors.

③ Tasting the tea: the deteriorated tea has a much less fresh or pure taste. For example, green tea may lose its original freshness, and black tea may become sour.

95. How is the new tea and aged tea distinguished?

Aged tea refers to the tea produced in the previous production year and before. Due to the long storage time, the contents of the tea are oxidized for a long time. As a result, the quality of the tea is reduced and the aroma is lightened. These changes help to identify the aged tea from the fresh tea. The fresh green tea is green,

lustrous, fragrant and mellow, and the color of the tea brew is clear. Yet aged tea is yellow and not as fresh as the new tea, and has a little scent. If the storage conditions are not good, the aged tea should have a stale taste and the infusion would be yellowish. The fresh black tea is black and bloom, fragrant and mellow. The infusion is red and shiny, and the steeped leaves are red and bright. However, the aged black tea in poor storage is dark in color with little aroma, bland or stale taste. The color of the infusion of the aged is dull and not as bright as fresh tea. And the aged steeped tea leaves would be in dark red and unable to stretch.

96. What are the best times for consuming different types of tea?

The best-before date varies for different types of tea. Non-fermented green tea, light-fermented southern Fujian Oolong tea, and Taiwan Oolong tea should be consumed during the year of production. For black tea, yellow tea, heavy fermented or baked Guangdong Oolong, northern Fujian Oolong, etc., it is recommended to drink them within 2 to 3 years after production. White tea and dark tea that need to go through the process of maturity transition can be stored for a period of time. The quality of Pu'er tea can be obviously improved after storage, so it is recommended to drink Pu'er tea after a period of storage. Take Pu'er tea as an example, raw Pu'er tea can be stored for 15-20 years under suitable conditions and the taste would be better. It is advisable to store post-fermented Pu'er tea for 5-8 years. It is not recommended to store the tea for as long as possible. After a long period of storage, the flavor substances in the tea are lost and the taste becomes pale. For example, the Qing Dynasty tribute tea in the Forbidden City would have only a mild aroma and taste after nearly two hundred years of storage.

97. How is the Pu'er tea stored?

One should be aware of the influence of oxygen content, temperature, light and humidity on Pu'er tea during storage.

① Air: Pu'er tea should be stored with circulating air but not in a draught. In addition, it should be protected from odors such as odors in kitchens and the industrial environment.

② Temperature: The storage temperature of Pu'er tea should be neither too high nor too low. It is advisable to store it at the local room temperature. It is not necessary to create an environment with a certain temperature. The normal indoor temperature, ideally between 20℃ and 30℃, would be suitable for the storage. Too high temperatures accelerate the fermentation and acidification of the tea.

③ Light: Light provides the energy that changes the chemical composition inside the tea. If the tea leaves are exposed to the sun for a day, the color and taste of the tea change significantly, thus losing its original flavor. Therefore, tea must be protected from light.

④ Humidity: An excessively dry environment can slow down the aging of Pu'er tea. In a dry environment, a small glass of water can be placed next to the tea to increase the humidity in the air. However, a too humid environment can lead to rapid changes in Pu'er tea. One of the common changes is "mildewing", which can render the Pu'er tea undrinkable. It is recommended to maintain the humidity at around 75%. In coastal areas, the humidity is higher than 75% in the rainy season due to the warm maritime climate. Thus, timely ventilation for distributing moisture is necessary during storage.

98. How is the white tea stored?

After storage, the color of white tea becomes darker and the taste of the brew becomes mellower and thus the tea is loved by many consumers. There are four requirements for the storage of white tea. First, if the white tea is stored for a long time, its moisture content should be maintained below 5%. Second, the storage environment should be dry, free of odor, and the humidity should be below 60%. If the room for storage has a cement floor, the soleplate should be supported by wooden frames. Third, white tea should be stored at low temperature and kept away from light since the chemical changes of the components in the tea and aging process are accelerated if the temperature is high and photochemical reactions of the components in the tea result in loss of the original flavor. Do not open the package during storage because the air entering into the package can accelerate the oxidation of the tea.

99. Does raw Pu'er tea become ripe Pu'er tea after storage?

Raw Pu'er tea does not become ripe Pu'er tea after storage under the natural condition. Raw Pu'er tea refers to the pressed tea made of sun-dried semi-finished tea leaves of Yunnan large-leaf (Daye) species. The color of the dried tea is blackish green. Ripe Pu'er tea is also made of sun-dried semi-finished tea leaves of Yunnan large-leaf species but already go through the process of pile-fermentation among other manual processing. The color of the ripe Pu'er tea is dark brown. In brief, ripe Pu'er tea is produced by manual processing of pile-fermentation, while raw Pu'er tea has not gone through this process. Raw Pu'er tea may have a change in quality during storage but it is a natural aging process. Raw Pu'er tea will not become ripe Pu'er tea in the end. They have a big difference in their taste.

100. How many materials are there for the package of tea?

Tea packaging refers to the containers or materials used to protect the quality, sanitation and safety of tea, extend its shelf life, and facilitate its transportation, weighing, preservation, display and sales. There are a variety of packages and materials for tea.

Packaging materials include cardboard, craft paper, white cardboard, plastic film, aluminum foil compound film, glass, etc. Containers include iron caddy, paper caddy, aluminum caddy, bamboo or wooden container, porcelain jar, tin caddy, etc. The packaging of tea plays an important role in keeping the quality of tea in the process of processing, marketing, storage and distribution, and preventing and reducing the deterioration of the color, aroma, taste and nutrition of the tea.